Mobile Communication

Digital Media and Society Series

New technologies are fundamentally altering the ways in which we communicate. This series from Polity aims to provide a set of books that make available for a broad readership cutting-edge research and thinking on digital media and their social contexts. Taken as a whole, the series will examine questions about the impact of network technology and digital media on society in all its facets, including economics, culture and politics.

Published:

Mobile Communication

RICH LING AND
JONATHAN DONNER

polity

First published in 2009 by Polity Press

Reprinted in 2010

Polity Press
65 Bridge Street
Cambridge CB2 1UR, UK.

Polity Press
350 Main Street
Malden, MA 02148, USA

ISBN-13: 978-0-7456-4413-4
ISBN-13: 978-0-7456-4414-1 (paperback)

A catalogue record for this book is available from the British Library.

Typeset in 10.25 on 13 pt FF Scala
by Servis Filmsetting Ltd, Stockport, Cheshire
Printed and bound in the United States by Odyssey Press Inc.,
Gonic, New Hampshire

The publisher has used its best endeavours to ensure that the URLs for external websites referred to in this book are correct and active at the time of going to press. However, the publisher has no responsibility for the websites and can make no guarantee that a site will remain live or that the content is or will remain appropriate.

Every effort has been made to trace all copyright holders, but if any have been inadvertently overlooked the publishers will be pleased to include any necessary credits in any subsequent reprint or edition.

For further information on Polity, visit our website: www.politybooks.com

Rich Ling:
To Dad, Grandaddy Seyler and Grandpa Ling

Jonathan Donner
To Calliope

Contents

Preface

To get a perspective on the rise of the mobile telephone, it is perhaps appropriate to start with an ode to the past, namely the eclipsing of the phone booth by the mobile phone. Indeed, the mobile phone is helping to push the telephone booth into history.

People born since the mid-1990s may never set foot in a phone booth. Thanks to mobile telephones, most of them have, or will soon have, the ability to reach anyone else they want to, regardless of either person's location. They will never have to deal with the search for an evasive phone booth in an unfamiliar location. They will not have to rummage madly for a dime (or a krone or a pound or a frank) to buy a few minutes of time. In short, they will not fully understand the way that the phone booth was a shared experience and cultural icon.

Future generations will never really understand that a phone booth was a place where hearts could swell or be broken ("Do you really want to go steady?"); where invitations were received and meetings arranged ("The party is at Frank's house? Great! We'll be there with the beer soon"); where important information was recovered ("What was Emil's address again, I wrote it down but I lost the piece of paper?"); and where deals could be done and undone.

The phone booth was also the location for different types of hi-jinks including prank calls, petty vandalism and phone booth stuffing (the record for stuffing a booth is variously reported as being 22, 24 and 25 persons). The phone booth played a role in a variety of films including Alfred Hitchcock's *The Birds* and Clint Eastwood's *Dirty Harry*. It is where the

stumbling spy Maxwell Smart enters into CONTROL headquarters and it is where Clark Kent somehow transforms
himself into Superman. (What did Clark do with the suit
of clothes he was wearing? Is it somehow returned to him
as though from the dry cleaners or did Superman leave his
suits piled up on the floor of different phone booths around
Metropolis?) On the darker side, the phone booth is a frequent
setting where movie murderers, blackmailers, kidnappers,
extortionists and all-round bad guys made anonymous,
untraceable calls outlining their demands or proclaiming
their – usually temporary – invincibility.

The phone booth was a symbol of temporary shelter, home-
lessness and a nomadic lifestyle. Bruce Springsteen talked about
sheltering himself in a phone booth on cold winter nights and
calling his girlfriend. In another case, the oddly placed Mojave
phone booth became a cult location – and phone number –
given its anomalous placement miles and miles from any major
road, building or normal semblance of civilization.

In their time, phone booths were a stylish bit of architec-
ture. The classic British phone booth was designed by the
architect Sir George Gilbert Scott who, when not working on
what were also referred to as "silence cabinets," found the
time to design the Liverpool Anglican Cathedral, Waterloo
Bridge, the library at the University of Cambridge and the
Bodleian Library at Oxford. Some suggest, rather morbidly,
that Scott's design was inspired by the tomb of Sir John Sloane
in St. Pancras' Gardens – to which it does, incidentally, bear a
passing resemblance.

The phone booth is the past. Since the mid-1980s, we
have seen the rise of a device that is slowly but surely replacing the phone booth, namely the mobile phone. The mobile
phone is becoming the locus of the calls that mark the different phases of our lives. Lovers are cooing to one another; bad
guys are making demands; teens are texting with their friends;
profit-driven young bucks are trying to move the markets; and
parents are trying to keep up with the delivery of children to a

spectrum of birthday parties, soccer practices and after-school engagements. Farmers in India are using the mobile phone to check the price of rice at the local markets, Filipina maids in Singapore are using it to send money home to their families, and entrepreneurs all around the world are buying, selling and arranging their affairs via the device.

The mobile telephone is also becoming a cultural icon in its own terms. The style, model and features of a phone all play into the image that we display to the world. Children can buy toy mobile phones or balloons in the shape of the device. They feature in films, and the release of an iPhone or the most recent *keitai* (the Japanese word for mobile phone) can make the headlines. The mobile phone is even generating its own form of offbeat contests, such as the mobile phone throwing competition (the current record seems to be 89 meters).

The mobile phone also has a much broader exposure than did the phone booth. The so-called "telephone ladies" (women who run a small independently owned mobile phone-based telephone service in Bangladeshi villages) provided, often for the first time, a local telephone service for their villages with a shared telephone link to the broader world. Whether formal or informal – just a lawn chair, an umbrella and some minutes to re-sell – "Public" mobile phones have spread further and faster than the public call offices (PCOs) offered by most landline companies in the developing world ever did or could. Ironically, as with the traditional telephone booth, the telephone lady in Bangladesh and the public mobile phone in Ghana are also being replaced by individual ownership of mobile phones. As we will see below, the private ownership of mobile phones is skyrocketing across the developing world, while in many more prosperous countries there are now more mobile phone subscriptions than there are people.

We are moving out of the box – the phone box – to what Goffman called "the un-boothed" phone. This book examines the incredibly rapid spread of the mobile phone around the world, and how we are adapting to its presence.

Acknowledgments

As with any piece of work, this book draws on the experience, insight and courtesy of a broad variety of people. We wish to thank colleagues and friends who have contributed to the writing of this book.

Leopoldina Fortunati, Casey Jenkins, Raul Pertierra, Jack Qiu, Marit Sandvik, Satomi Sugiyama, Carolyn Wei and Rajesh Veeraraghavan have helped us with the development of the fictional vignettes. Their grounded knowledge of the different milieus was essential in helping us render these situations. Our extended network of colleagues, including James Katz, Richard Chalfen and Scott Campbell, have been supportive in their various ways

At Telenor, Nisar Bashir, Christian Nøkleby, Per Helmersen and Hanne Cecilie Geirbo have provided help and insight, each in their own way. At Microsoft Research India, Kentaro Toyama and P. Anandan have done the same. In addition, the California-based Tom Farley is, as always, a valuable resource when it comes to the history of the mobile phone.

CHAPTER ONE

Introduction: the quarter-century beyond the Maitland Commission Report

1 Introduction

In 1982, a conference with the imposing name of "The Plenipotentiary Conference of the International Telecommunication Union" formed a commission to take up the question of access to telephones in the developing world. Two years later, that commission, chaired by Sir Donald Maitland, issued its report with the equally imposing title, *The Missing Link: Report of the Independent Commission for Worldwide Telecommunications Development*.

What many policymakers today call simply "The Maitland Report" outlined the impact of telecommunications on the effective operation of public service, commerce, health services, agriculture, banking services, etc. (Maitland, 1984, 7), and examined how telecommunication facilitates coordination and makes transport systems more effective. But it is best remembered for its stark, sweeping statistics describing the discrepancies in telecommunication services between the developed and developing worlds. The report noted that more than 50 percent of the world's population then lived in countries with less than 1 telephone for every 100 people, and that many of those telephones belonged to offices and businesses, out of reach of everyday citizens. In many countries, it argued, there was literally no telecommunications service outside the more populated towns and cities. Writing in an era before the widespread use of the internet and mobile telephones, the commission lamented: "More than half the world's population live in countries with fewer than 10 million telephones between them and most of these are in the main cities; two-thirds of the world's population have no access to telephone services. Tokyo has more telephones than the whole of the African continent, with its population of 500 million people" (Maitland, 1984, 13).

The commission was not blind to the march of technological innovations occurring in global telecommunications at the time. Its report discusses the possibility of using radio in lieu of wired landlines in the service of supplying telephony. It saw microwave systems as an alternative to long-distance trunk lines, satellite systems as serious alternatives for the provision of telephony to rural areas, and terrestrial radio as a way to extend telephony's reach. In the only real mention of what would become the mobile telephone system, the report's authors argue: "Improving the effective utilization of the frequency spectrum is possible by using the cellular concept and other methods of dynamic frequency assignment" (Maitland, 1984, 31). However, seen with a quarter-century's remove, it is clear that the Maitland Commission largely failed to predict the role cellular-based mobile communication would play in revolutionizing the accessibility of telecommunications around the world.[1]

To be fair, when the members of the Maitland Commission were at work on their report, mobile telephony was still in its infancy – not yet even considered a yuppie plaything. Experiments in 1969 had used a cellular system to place calls on an Amtrak train traveling between New York City and Washington, DC. Only a few years later, in 1973, Martin Cooper placed the first commercial call on a handheld mobile phone in New York. However, in 1984, the commercialization of the Global System for Mobile Communication (GSM) – the form of mobile communication that is most widely adopted – was still a decade in the future.

This, the picture painted by the Maitland Report, in many ways serves as a baseline against which to measure the digital telecommunications revolution that has followed. We have seen a dramatic change in our access to telephony, and in particular to mobile communication. Inexpensive and used handsets have made it possible for more of the world to partake of these signals. According to the International Telecommunications Union, which keeps track of these things, there were 3.3 billion mobile subscriptions by the end of 2007 – approximately 1 for every second person in the world (ITU, 2008a) – and recent figures suggest the 4 billionth subscription became active at the end of 2008 (Cellular News, 2008; ITU, 2008b).

Unlike landlines, these billions of mobile handsets rarely hang on walls or sit on office desks. Instead, mobiles belong most commonly to individuals, who can carry one or more of them around wherever they go. As we will discuss in more detail in the pages that follow, mobile phone users are more reachable than ever before, and can reach out to others more easily than ever before. If the explosion in *connectivity* is the first major theme of the mobile boom, then this new level of *reachability* is the second.

This dramatic explosion in mobile telephony since the publication of the Maitland Report is a one-time occurrence. We will never see it again. In this book, we wish to examine the consequences of that change and to explore how it is working its way through society. We will look at how it affects the lives of people from around the world and how academicians are trying to describe its impact. We will look at the positive as well as the negative impacts of the mobile phone. Finally, we will examine how the mobile phone provides us for the first time with a mediated form of individual addressability.

Transitions such as this are rare, yet it is in such moments that the interaction – or should we say clash – between the technology and the accepted way of doing things gives us a chance to see the inner functioning of society. To put this into a sociological perspective, it is as though we are experiencing a type of breaching experiment (Garfinkel, 1967). Those who are familiar with Garfinkel's breaching experiments know that they were designed to give us the opportunity to see how we respond when the rules of society are changed. The unprecedented altering of expectations shows us how people understand the unforeseen situation and how they patch together a way of making sense.

In many respects, common questions and tensions surrounding mobile communication use allow us to take stock of what constitutes appropriate social interaction. Do mobile phones encourage political protest? Do we need to respect the sanctity of time-based appointments since we can easily renegotiate them as needed? What are our feelings towards free communication between teens? Should we always be available to everybody? What attentions do we owe co-present interaction when competing with the allures of talking with a friend?

The transition from a world of landlines to a world of mobiles provides us with a unique chance to gain insight into how a personal technology affects social organization, both for the better and for the worse. After the boom is complete and mobile use becomes the worldwide norm, it will become harder once again to discern these interactions between mobile communication and society. If we squander this chance to study mobile use, it will not come again.

2 Increased connectivity: mobiles sweep the world

The first decade of the twenty-first century may be remembered as the historical moment when the majority of the world's population first secured easy and affordable access to telephones. Of course, the telephone was invented in the nineteenth century, and it steadily gained in popularity and complexity throughout the twentieth century. In fact, between 1976 and 2000, the number of landlines in use nearly quadrupled. By the turn of the millennium, there were 1.7 billion telephones on the planet: 983 million landlines, and 740 million mobiles.

While the PC and the internet have received much attention, it is the mobile telephone that has enjoyed a quick and

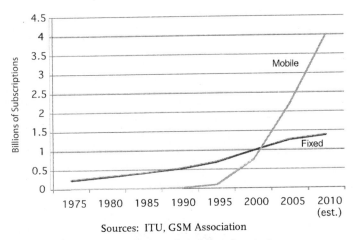

Sources: ITU, GSM Association

Figure 1.1 Worldwide landline and mobile subscriptions, 1975–2010 (est.)

broad level of adoption. In 2006, there were about 10 internet and 32 mobile phone users per 100 persons in the world (ITU, 2008a). The telecommunications story of the first decade of the twenty-first century is undoubtedly mobile. Between 2001 and 2010 the planet will probably add another 400 million landlines, and a staggering 3 *billion* new mobile subscriptions. If these estimates about the near future are to be believed, there may be over 5.4 billion telephone subscriptions on the planet in 2010, 1.4 billion landlines and at least 4 billion mobile subscriptions (Cellular News, 2008; GSM Association, 2008; ITU, 2008b).

We stress "may" at this point, because both the historical and projected estimates in figure 1.1 are aggregations, compromises and best guesses (James & Versteeg, 2007; Sutherland, 2008). The staff of the International Telecommunication Union, a UN body in Geneva, has gathered estimates by querying member states and telecommunication companies (or "telecoms") from around the world, who in turn have gathered their estimates by polling individual operators and national regulators. They have settled on subscriptions, in overall and

per-person terms, as a common mode of measuring and com-
paring mobile use around the world and over time.

Counting subscriptions, however, can be problematic since
many users are missed in the count. Some users will take
advantage of formal mobile payphones that are run as businesses (Aminuzzaman, Baldersheim & Jamil, 2003; Bayes,
2001), while others use informal ones where payment can
be in the form of reciprocity (Sey, 2006). Countless others
will share phones within a family (Chavan, 2007; Goodman,
2005; Konkka, 2003; Samuel, Shah & Hadingham, 2005), to
place and receive calls even if they cannot afford a handset
or subscription of their own. In some cases, subscribers purchase only the SIM card[2] and borrow the handset of a friend
or neighbor until they can afford a handset. Looking at the
situation in Asia, Zainudeen et al. (2007) found that, across
four countries with relatively low aggregate mobile phone
penetration (Pakistan, Thailand, the Philippines and Sri
Lanka), something over 90 percent of the respondents had
used a phone in the previous three months. They reported
that 80 percent of the respondents were within a five-minute
walk to the nearest telephone.

In other ways, counting subscriptions can overestimate the
number of people with a subscription, or the number of active
subscriptions. We have seen numerous countries surpass
100 percent in terms of penetration. For example, in 2006
the ITU reported that Norway had 108.6 subscriptions per
100 persons. Yet during the same period, data from a random
sample of 1,000 Norwegians aged 13 or older indicated that
92.8 percent "Owned their own mobile phone."[3] When asked,
about 7 percent of the teen and adult population said that they
did not have a mobile phone. If there are 108.6 subscriptions
per 100 persons but only 92.8 percent of the people over 13
say they have a mobile phone, something is clearly awry with
the numbers. Apart from "dead" subscriptions (subscriptions
that are still on the telecom's books but which the subscriber
no longer uses), it is clear that some people have more than

one subscription. In the year 2000, about 13 percent of teens in Norway reported that they had two or more telephone subscriptions (Ling, 2004). The motivation is often to save money by "SIM switching," that is, for example, selecting the subscription that is the least expensive at a given time during the day. Similarly, although many prepay SIMs are discarded by users, operators sometimes carry these dead and dying subscriptions on their books for months or years after users have given them up.

Thus, while estimates suggest that by 2010 there will be at least 5.4 billion fixed and mobile subscriptions on the planet, this does not mean that 5.4 billion people own a telephone. Nevertheless, we can get closer to a back-of-the-envelope estimate of the number of telephone owners in a couple of steps. First, we discount the mobile figures for over-counting subscribers. Sutherland (2008) suggests that, of the 3.3 billion mobile subscriptions active at the end of 2007, per-haps 500,000 should be removed as over-counts. Keeping that same proportionality, if we discount the newer 4 billion mark by the same proportion Sutherland uses, we would arrive at an estimate of 3.4 billion active mobile subscriptions. Second, we can discount the mobile figures by contrasting those who have a mobile subscription and a landline subscription, and those who have only a mobile subscription. Hamilton (2003) contrasts mobile phones as *complements* (among users who add a mobile line to a fixed line) and as *substitutes* (among those who have only a mobile). In prosperous countries, some people, particularly youth, are giving up landlines by choice (Blumberg & Luke, 2007). For hundreds of millions of other users, particularly in the developing world, the mobile is the only affordable option. With this distinction in place, we can isolate the mobiles used as substitutes for landlines as the ones that add to the proportion of the world's population that has a telephone. To be conservative, let's assume that in 2010 there will be 1.4 billion landlines and 3.4 billion active mobile sub-scriptions. Let's further presume that every single landline in the world is paired with a complementary mobile – that the first 1.4 billion mobile subscriptions sold do nothing at all to raise the proportion of the world's population owning a telephone. That would still leave 2 billion unpaired, substitutive mobile subscriptions in 2010. Roughly speaking, 2 billion "mobile only" subscribers and 1.4 billion "landline plus mobile" subscribers, would sum to 3.4 billion subscribers . . . just about "half the world" of 6.8 billion people by 2010.

Until we have worldwide surveys counting users rather than subscriptions, our numbers will remain approximate. However, the relatively recent arrival, and even more recent sudden uptake, of mobile use around the world is staggering. The arrival of mobiles has, in one decade, roughly tripled the total number of ways to connect to the world's telecommunication grid (so too have PCs and internet telephony, but that's a story for another day). At the same time, perhaps 2 billion people acquired their first telephone during the decade.

Most of these new first-time telephone owners will be mobile owners, and most of them are in the developing world. As of 2008, 58 percent of the world's mobile phones are in these countries (UNCTAD, 2008). As figure 1.2 illustrates, China is the world's largest mobile market; India will soon overtake the USA as the second largest (IE Market Research, 2008). Figure 1.2 also illustrates that, even with this recent surge in mobile use, mobile subscriptions are not evenly distributed across the globe. Of the ten countries with the most mobile subscribers in 2007, four were from large prosperous nations in the Global North, five were large nations in the developing world (Global South), and one, Russia, is a transitional economy. Although the gap is closing, the prosperous nations have higher penetrations (use per 100 citizens) of mobile subscriptions than poorer nations. Like Norway, mentioned above, Germany and Italy (shown on figure 1.2) each have more mobile subscriptions than people; India had the lowest penetration among the top-ten markets with less than 20 mobiles per 100 persons. Eritrea (1.44 per 100) and the tiny islands of Kiribati (0.75 per

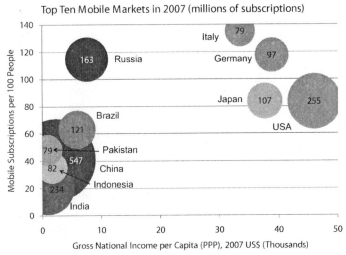

Top Ten Mobile Markets in 2007 (millions of subscriptions)

Sources: ITU ICT-Eye, http://www.itu.int/ITU-D/icteye/,
GNI from World Bank Development Indicators,
www.worldbank.org/data/dataquery.html

Figure 1.2 Where the mobiles are, 2007

100) had the lowest penetrations among all countries report-ing data to the ITU in 2007.

Why the boom? It is usually less expensive to install and maintain a cellular tower to serve a neighborhood or village than it is to bring the necessary landline cables across landscape into individual households. This fundamental difference has changed the coverage–access–ownership equation (Donner, 2005; Dymond & Oestmann, 2003) for millions of people who were unable to own a landline, or even in some cases walk to a payphone. Indeed the World Bank estimates that, by 2005, 77 percent of the world's population lived under a mobile signal (World Bank Global ICT Department, 2005).

This increase in what is known as "teledensity" has social impacts in more prosperous parts of the world, as well. In one such area, namely Scandinavia, material from the national census bureau in Norway revealed that, quite literally, all 15-year-olds had a mobile phone.[4] That is, the national census bureau with their advanced forms of random selection and their exhaustive survey recruitment techniques were not able to find 15-year-olds who did not have a telephone. This, too, is a profound change, one with social implications for family dynamics, for public safety and public protest, for work–life balance and for the meaning of adolescence itself. We will revisit some of these questions in the next section, and later in the book.

3 Increased reachability: social consequences of the mobile phone

The mobile telephone changes the way we communicate with one another. Instead of calling to a fixed geographical location as is the practice with the landline device, we call to an individual, wherever they may be. It also allows us to interlace our telephonic communications (and our text messages) into the weave of our other activities.

The mobile phone has developed into a type of safety link for those who would otherwise be tied to physical locations. It has changed the way that we coordinate, or perhaps microcoordinate, our meetings and our daily interactions. It has become the de rigueur accessory and it has given rise to the practice of text messaging. It has changed the way that teens interact with parents and with peers and it has changed the dynamics of social networks and the development of social cohesion.

In a very short time, the device has had a major impact on the way we interact and organize our lives. Mobile telephones are used by teens to make and break appointments and to keep their friends updated. Texting, a concept that barely existed a decade ago, is used by teens and interestingly also by deaf persons to communicate (Bakken, 2005). It can be used by students to keep track of their social network while they are in class and it can be used by groupies to follow the movements of celebrities such as the UK's Prince William or, for those with a more decadent bent, Paris Hilton.

Lovers use the mobile to exchange endearments. Soccer Moms (and Dads) coordinate their kid's next pick-up. Protestors use mobile to outwit police and rattle governments. In the developing world it has spawned "telephone ladies" in Bangladesh (Singhal, Svenkerud & Flydal, 2002) and "umbrella ladies"[5] in Ghana (Sey, 2006), not to mention texting matchmakers in the Philippines (Ellwood-Clayton, 2003). Fishermen in the Indian state of Kerala use the device to find the best price for their daily catch (Jensen, 2007). Stressed-out small business owners around the world grab their handsets to take orders and find suppliers (Donner, 2005). Many workers, formerly desk-bound, can carry out their jobs while away from the office or even during leisure time – for example, you could be a poolside real estate sales person. The mobile phone is also used in the pursuit of "alternative" and more seedy business transactions such as prostitution and narcotics sales. In short, billions of people in every country on earth use the mobile to arrange important (and not so important) affairs. No task seems to be too large or too small for the mobile phone.

The actual mobile telephone – that is, the physical handset that we carry with us – can be as simple as a small plastic device with few functions beyond the ability to send and receive calls. At the other end of the scale, they can be elaborate gold and jewel-encrusted pieces of wearable art. They can be a $15 recycled or bootlegged device bought from a small stall in a dusty market in Bangalore, or can be powerful "smartphones" allowing for a spectrum of communication and data-related functions. Some call it the "Swiss Army knife" (Jenkins, 2006) of technologies since, in addition to being a communications device, it is becoming a camera, a photo album, a rolodex, an e-mail reader and a calendar. There are also devices that provide location information, and can also function as a small-change purse, a bus ticket and a mini-notebook.

It is not strange that the device has gained an iconic status. There are balloons shaped as mobile telephones; small children play with and even make their own "dummy" mobile telephones;

Steve Jobs created a sensation by introducing the iPhone. In Ghana, it is possible to buy fantasy coffins made in the form of a mobile telephone. Thus, it is not simply a functional device, but has entered into our symbolic panoply (Katz, 2006).

4 Theoretical lenses

For millions of people who have lived without telephony, mobiles provide basic connectivity: a way to gather information and to be in contact with the world. Further, mobile communication makes each of us directly addressable, regardless of where we may be. It allows for the interlacing of different activities. We can attend a lecture and at the same time arrange via SMS (Short Message Service) to meet a friend for coffee afterwards. In these small but persistent ways, it supports the development of social cohesion among persons in the intimate sphere and it gives us the means to control interactions in ways that were not possible before. For these reasons, it is clear that the mobile phone plays an increasingly important and central role in many people's daily lives, and has consequences for the societies we live in.

There are several relevant theoretical lenses we can use to account for and understand many of these changes. Some scholars have applied existing theories of communication technologies to the "new" case of the mobile phone, without fundamentally altering those broader theoretical lenses. Others have developed theories more specific to the mobile. We will review some of the more notable theoretical applications below. However, given the broad ranges of disciplines, methodologies and theoretical perspectives of the researchers involved, we do not expect a single, integrated "theory of mobile communications" to comprehensively cover all the issues at hand. We assert that the diversity of interpretations of the mobile's overall impact and interactions with society are driven as much by the diversity of theories at work as by the complexity of mobile use itself.

In fact, even the question of what a "mobile phone" or "mobile communication" is is a matter of some debate. People make voice calls on mobiles, which opens the technology for comparison with exiting assessments of landline telephones (de Sola Pool, 1977; Fischer, 1992). People send text messages, which, as an asynchronous, peer-to-peer practice might have some attributes in common with e-mail or instant messaging (IM), but might be better served by fresh assessments of the text message as a medium unto itself (Harper, Palen & Taylor, 2005). Beyond that, the sky's the limit. Mobile web browsers beg to be understood vis-à-vis traditional web experiences. Camera phones open one to theories of popular photography (Chalfen, 1987); embedded MP3 players might require a look at the research on portable media players and the original Walkman (Bull, 2001). Even the tiny "occasional games" might be understood by drawing on broader theories of games and society (Rheingold, 2002).

At the same time, the "Swiss Army knife" that is the modern mobile phone sets up a conundrum of convergence – the modern mobile phone is a central actor in "convergence culture," in which various media intermingle across multiple platforms, and are consumed, re-cut, republished and re-appropriated by active users in unpredictable ways (Jenkins, 2006). We'll return to some of these convergence themes later on. However, for the most part, we have elected to keep the primary focus of both this overview of theory and the book as a whole on the practices of voice calls and text messages. While the additional functions are useful in many situations – and are fascinating to marketers and theoreticians alike – we start from the assertion that most people use the device most of the time to talk and text to one another.

4.1 Applications of broader theories to the case of the mobile

Since the mid-1990s, many existing, broader theoretical perspectives have been brought to bear to understand the adoption and use of mobile communication. These cover a full range of important questions, from identifying the drivers of adoption to the dimensions of impact on individuals and societies; from describing micro-level use in all its complexity to refining broad general theories of the nature of information technologies in the production and maintenance of a global social order (Donner, 2008a). We'll review a few of these theories in the pages that follow.

4.1.1 ADOPTION

Some questions naturally resurface whenever a new technology arrives on the scene: *Who chooses to adopt it? Under what conditions? And why?* For decades, this approach has been almost synonymous with Rogers' theory of the diffusion of innovations (Rogers, 1995). The application of Rogers' approach to the mobile phone is particularly interesting, since in many ways, for the general population, the device represented a previously unknown method of using the telephone. Rogers has carefully analyzed how new innovations are adopted by society. He describes how innovators and then early adopters, etc., each adopt a particular innovation in their turn. Each group is motivated by different things. Innovators are the people who are perhaps most guided by an inner compass. They have a need to be the first to use a device or an innovation. They are attuned to the information sources that commonly announce the appearance of novel developments. They are willing to take chances and they are willing to suffer through the various teething problems of new technologies. These were the individuals who bought the large "lunch-box"-sized mobile phones and started to explore their use. Based on the experiences of the innovators, the next group, namely the early adopters, start using the product or service. The early adopters base their decision to adopt on the experiences of the innovators. They are more patient and they use interpersonal interaction to gather information more than the innovators do. This group, however, is a very important one, in the sense that they legitimate the adoption process. When a particular development has survived the critique of the innovators, and enough of the rough edges have been knocked off of the adoption process, the early adopters start to use it. This is an important signal for the groups that follow. The adoption of the mobile phone by these central individuals sent a message to others that the device was not just for specially interested nerds, but had potential for broader use. Following the early adopters come what Rogers called "the early majority," the late majority and finally there are the laggards. These later groups adopt communication devices when there is a critical mass of users.

The issue of critical mass is a special dynamic associated with the adoption of communications innovations (technologies designed to allow point-to-point communication, such as the fax, the mobile phone or social networking software such as Facebook) that is not seen with the adoption of stand-alone innovations such as MP3 players, digital cameras, etc. This is perhaps best seen with the adoption process of the fax machine. If there are only a limited number of fax machines to which you can send faxes, the device is of little use. However, the value of your fax machine indirectly increases with the addition of each additional fax machine to the universe of devices, since it increases the chance that your interlocutor will have access to a fax machine should you need to send him or her a message. As the number of mobile phones (or fax machines, or Facebook subscriptions) increases, they collectively become more used as an accepted form of communication and thus it is more difficult to NOT be a user. It is clear that some technologies can be superseded by others. The fax machine was supplanted by e-mail, IM and social networking and thus is little used today, just as the telex machine before it was supplanted by the fax. These technologies nonetheless illustrate the dynamics of innovation and critical mass. Rogers suggests some of the key issues during this process of adoption; however, his analyses seemingly stop upon the purchase of the item.

Designers have often focused on the affordances perspective (Gaver, 1991; Gibsen, 1979; Hutchby, 2001; Norman,

1990; Sellen & Harper, 2002). This approach examines how the characteristics of an innovation relate to the way we see using it. This type of insight is useful when designing new products. It does not necessarily, however, help in understanding the broader social consequences.

4.1.2 IMPACT

One of the more common theoretical tools for understanding technology in society is the idea of technical determinism – it is also one of the more common straw horses that gets beaten among social scientists. Basically, this says that as a new technology is introduced it reformulates society in its image (Cottrell, 1945; Mumford, 1963; Sharp, 1952). The classic example of this is the Marxian notion of technology determining the form of society. Another example of more technically based theories is that of Beniger and his notion of the control revolution (see chapter 6). According to Beniger (1986) there has been a progressive increase in the speed of production processes. This has been accompanied by the need for control processes. As industrial production increased in speed, it became obvious that information processing was needed to control manufacturing. As trains increased their speed and factories spat out goods at faster rates, there was a corresponding need to route the trains and to control the logistics systems. Beniger suggests that this is the broad motivation behind the development of ICTs. He focuses on the issue of industrial production. The control technologies spawned in that arena have in many cases also been used for social as well as commercial purposes. The telegraph, the landline telephone and later the integrated circuit, with its legacy of computers and mobile phones, have all spilled over into the non-commercial sphere.

A major theoretical direction in communication research is the so-called "media effects" approach, which focuses on how mass media influence audience attitudes and behavior (see, for example, McCombs & Shaw, 1972, on the agenda setting function of the media). The assertion here is that the media (TV, film, radio programs) are the active partners and that the viewer is more the passive recipient of the media's influence. The study of mobile communication clearly strains in order to apply this approach. In the case of mobile communication, the interaction is more often between two interlocutors who are each responsible for producing their own "content" in the form of contributing to the conversation. The question of media effect then becomes not the impact of a relatively powerful content producer and a relatively powerless consumer of this content, but rather the impact of the mediation system on the ongoing interaction between the conversants.

4.1.3 USE

An alternative to the deterministic impact models is the social shaping approach, which suggests technologies arise out of social contexts and the technologies that are brought forth are modified according to the needs of the users (Bijker, Hughes & Pinch, 1987). We see that the mobile phone was engineered with certain assumptions about how it might be used. This included services like caller ID and (with the GSM system) the separation of handset manufacture from the operation of the telecom networks via the use of the SIM card. These features have been used in a variety of ways, including ones that were not envisioned by the people involved in the specification of the system. Developers perhaps imagined serious business-related uses of mobile communication, while use has come to include giggling teens as well as people who are only on the margin of the money economy.

Both the technical determinism and the social shaping approaches are useful for understanding the interaction between technology and society, but each has its baggage. Technical determinism focuses on the brilliant and perhaps individualistic actions of the inventors. It does not allow for the role of the broader social drift in society. In the social shaping approach, the interpretation of a technology is seemingly open. It is the users who decide how a technology is applied. The mobile phone can, if pressed into service, be a flashlight or it can be a bookmark. The uses of the object are seemingly completely open.

If we are interested in examining the social consequences of mobile communication, as we are, another approach is fruitful, namely the domestication approach. Domestication examines how we go through the process of discovering, purchasing and integrating artifacts and services into our daily life. In addition, it provides a social dimension to understanding the expectations of use. Finally, domestication helps to account for how we judge others' use of artifacts (Haddon, 2003; Scifo, 2005; Silverstone & Haddon, 1996).

Domestication examines how an item or a service enters our everyday life. Our adoption of the mobile phone, for example, often includes a period of considering our actual need for a device and, upon deciding that we need to have a mobile phone, which device would be appropriate. Upon purchasing a mobile phone, the domestication approach helps us to understand the process of deciding when and how we use the device. Thus, unlike the adoption approach of Rogers, the analysis does not stop with the purchase of the artifact, but rather follows the process of placing the object or the service into the broader context of our daily lives. In addition, it does not focus directly on the issues associated with technical determinism or social shaping. There is not a sense that technologies automatically reformulate our lives in their images, or that the technologies are extremely elastic in the ways that they can be used. Rather, domestication says that the truth lies somewhere in between. There are advantages to technologies that cause us to adopt them, and upon adoption they have social consequences – for example, in the case of the mobile telephone, it provides basic connectivity, direct addressability, interlacing of activities, coordination and control, and social cohesion. In addition, domestication suggests that we redefine and adjust our uses of the technology according to our needs. These needs are not infinitely flexible. Using a mobile phone as a shoehorn, for example, does not really work. However, within some boundaries, the technology can be re-applied to new situations through the inventiveness of the users. One final dimension of domestication is that it helps to take account of how the adoption and use of an artifact colors others' estimations of the user. In this way, the domestication approach provides a backdrop with which to understand both the adoption and social consequences of mobile communication. To develop the metaphor, it also helps to understand the taming of the technology, its placement in our lives and finally the way that ownership and use of the object become elements in others' understanding of who we are.

It is not surprising that researchers in the cultural studies and anthropological traditions have brought a somewhat different perspective to the analysis of the mobile. These works often approach the mobile as artifact: a physical object, yes, but one constructed over time and interpreted, appropriated, and understood by communities in diverse ways, depending on social context, private need, and public display and observation. Recent examples of this perspective include Goggin (2006), who projects current use well forward into the problems and promises of 3G and 4G, of image and of video, as the mobile becomes a medium unto itself, and Horst and Miller (2006), who report on mobiles' use by low-income people in Jamaica. Their work illustrates the ways in which Jamaicans' evolving choices about how and when to use their mobiles are both reflections of and influences on the norms of communication, expression, sociality and, indeed, economic survival in the particular contexts that are rural and urban Jamaica. To Horst and Miller there is "no fixed thing called a cell phone or fixed group called Jamaicans" (p. 7), but there are a myriad of changing constructions, strategies and locally relevant meanings which emerge as people and the technologies interact over time.

If deep case studies like Horst and Miller's emphasize locality and regional specificities in mobile use, other sociological approaches take a broader view, linking the mobile to universal trends in the constitution of society. A recent and particularly notable example of this is the work by Castells, Fernández-Ardèvol, Qiu and Sey (2007). Through examples drawn from around the globe, they identify the "relentless connectivity" (p. 248), offered by the device as a source of a coherent "network logic" (p. 6). In turn, this network logic drives a range of social transformations related to mobile use, including: a growth in individual autonomy, particularly for teens; increased opportunities to build "networks of choice" and new communities; and the breaking down of the dichotomies between the producers and consumers of information. We will touch on many of the same transformations in the pages ahead; however, we close the section on "existing theories" of mobile communication with this review work because it is explicitly built on Castells' broader conceptualizations of the network society (Castells, 1996). Mobiles do not up-end the logic of the network society, built on the flows of information across the planet. Rather, "the mobile communication society deepens and diffuses the network society, which came into existence in the past two decades" (Castells et al., 2007, 258). While locally constructed and infinitely variable at the micro level, at this most macro of levels, argue Castells and his colleagues, mobiles ultimately play a role similar to that of other ICTs.

A pleasure and frustration of doing research at the intersection of humans and technologies, of disciplines and methodologies, is that it is possible to find both Horst and Miller and Castells et al. to be helpful and insightful, even where they clearly differ in perspective and conclusion. The specificity of the local context seen in the work of Horst and Miller and Castells et al.'s generalization of global integration can be complementary, not contradictory.

4.2 Newer mobile-specific theories

At the risk of further complicating matters, we now must turn to a discussion of newer theoretical approaches that have been spawned specifically while studying mobile communication. These new theories reflect a sense, shared by many of us, that mobile communication is "different" – different enough from, say, landlines or personal computers or pocket cameras – to require an explanation of its own.

In addition, mobile communication is a phenomenon that changes the dynamics of social interaction. In this process, it also exposes and challenges our assumptions about how social interactions should be. Where, prior to the widespread adoption of mobile phones, we chatted with our dining partners in a restaurant, now it is possible – and indeed somewhat common – to put those who are co-present on hold in order to have a telephone conversation. Thus, mobile communication exposes our taken-for-granted assumptions as to how we ought to interact in different situations. We illustrate just a few of these new theories in the next few paragraphs.

Christian Licoppe and his colleagues have examined the way that mobile telephony changes the frequency of interaction between close family and friends. They have suggested the concept of "connected presence" to describe this (Licoppe, 2004). Using data from France, Licoppe suggests that, prior to the advent of the mobile phone, we used the landline telephone to make longer calls that often took on the nature of a special occasion. Indeed, the interlocutors often set aside the time to have a good chat. They might have had a specific reason for calling, but the conversation could have covered many different themes and could have lasted for an extended period. The conversations could even have been seen as a routine part of the week, where the two conversation partners called one another for a chat every Saturday morning, for example. In this regime the partners might collect events and thoughts that they would discuss in the call. They might think that a particular situation was worthy of relating in the call, or recall that they were reminded of their interlocutor when they saw a particular item in a store or were confronted with a particular situation. The items on this mental list might or might not be discussed, depending on the turns and twists of the conversation.

With the adoption of the mobile phone, the threshold for interacting has been lowered. Among the inner group of family and friends we are, in effect, continually available to one another. There is not the sense that the call is a special occasion. Licoppe describes a situation where we do not have a single longer call that rambles from one topic to another. Rather we are engaged in a series of shorter communications (both voice and text) that are, in effect, an ongoing dialogue that lasts through the day. Thus, as things arise, the two conversation partners might call or text one another either to alert them to issues that are immediately relevant or to simply let the other know that they are in their thoughts. Rather than having a two-hour conversation in the evening via the landline phone, the two individuals maintain a steady exchange of patter. This is what Licoppe calls "connected presence." He goes on to describe it by saying that it:

> consists of short, frequent calls, the content of which is sometimes secondary to the fact of calling. The continuous nature of this flow of irregular interaction helps to maintain the feeling of a permanent connection, an impression that the link can be activated at any time and that one can thus experience the other's engagement in the relationship at any time. (2004, 141)

The mobile phone is particularly well adapted to this form of interaction since we always have it with us and we often have the numbers of our closest friends and family entered into the phone (Licoppe, 2004). In addition, access to texting via the mobile phone facilitates this form of interaction since it is relatively easy to produce a quick note that is not as intrusive as a voice call but still can communicate either a bit of information or perhaps simply a quick phatic point of contact. According to Licoppe, the calls and texts have become so engrained in our everyday lives that they are no longer worthy of note. They are simply an ongoing form of dialogue, albeit with remotely located members of our inner sphere.

In this way, the mobile phone is causing us to weigh up the importance of co-located interaction vs. maintaining contact with close family and friends regardless of where they might be. Rules of etiquette, for example, are often calculated to privilege those with whom we are physically co-present. Given Licoppe's connected presence, however, we might be more concerned with the wellbeing of our remotely located friend and this will play out in our lack of concern for our geographically immediate situation. We are clearly still in the process of working out how to deal with these gaps and inconsistencies. Licoppe's insight helps us to understand the issues that are being dealt with.

Another person working in this area is Leopoldina Fortunati. In her work, she examines the mobile telephone as a symbolic device and, more particularly, the mobile phone as a fashion item. She has written extensively on how the mobile phone has become, in many cases, an object in which the users invest a sense of identity, and how it also becomes a symbol of their personality (Fortunati, 2005c). This analysis suggests that, in addition to being a functional device, the mobile phone can also be viewed as a type of fashion accessory. Just as with other such accessories, it is selected and used with an eye towards the impression that it will give off. A large multi-functional PDA (Personal Digital Assistant) device perhaps describes the business guy; an iPhone might characterize a young urban type; the small intensely colored flip phone is the sign of an upwardly mobile fashion-conscious woman, etc. The codex of phones and their meaning is dynamic. That which was cutting-edge yesterday will likely be dowdy tomorrow. Fortunati's work helps us to focus on this dimension of mobile phone consumption and use.

Rich Ling (one of the co-authors of this book) has added the idea of micro-coordination to the discussion of mobile telephony (this idea is also discussed in chapter 4). Ling notes that there are several major social consequences that have arisen from the adoption and use of the mobile phone. These include an altered sense of safety and security, and changes in the way that youth experience the emancipation process. However, perhaps the major consequence of mobile communication is the way in which we use it to coordinate, and indeed micro-coordinate, the flux of our daily interactions with our closest friends and family.

Prior to the adoption of the mobile phone we often agreed to meet by negotiating a specific time and place. If we were late, if we lost our way or if we somehow misplaced the information about the time and the place, it was often difficult to get word to the people we were meeting. The mobile telephone allows for a type of nuanced and fine-grained interaction between individuals within the intimate sphere (family and friends). This connectivity means that we are able to coordinate our interactions at a greater level of specificity than before (Ling & Yttri, 2002). Another person who has suggested a similar effect is Anthony Townsend and his idea of the "real-time" city (Townsend, 2000). Looking at this somewhat broadly, the rise of large cities, complex and specialized production, and differentiated forms of interaction mean that we often need to attend to a variety of individuals and institutions in the course of our daily lives. We deliver the children to day-care in one location, work in another, shop for food in one place, and also need to visit other types of shops and commercial outlets for other wares. We need to coordinate the activities of children and other family members, each of whom has their own unique bevy of places they have to be and things that they have to do. The mobile phone facilitates this innately complex situation by giving ubiquitous accessibility to our near family and close friends.

Another area of analysis for Ling has been the study of the mobile phone and social cohesion. In his book *New Tech, New Ties*, he examines the mobile phone as a means through which we can perform social rituals that result in social cohesion, such as those suggested by Durkheim (1995), Goffman (1967) and Collins (2004). It is obvious that the mobile telephone facilitates the instrumental coordination of everyday life. At the same time, it is important to note that it provides us with a method of attending to the expressive needs of the intimate sphere. We can tell jokes, have lighthearted banter. We can gossip and we can engage in the texting style of our local group. According to Goffman (1967), it is through the use of these rituals that we develop and maintain social cohesion within the small group. These larger and smaller rituals (that can be co-present or in some cases mediated via the mobile phone) are the way that we are drawn into a common sense of group identity. A ritual might be a more distinct event such as a wedding ceremony, or something more casual such as simply sharing a few minutes of jovial conversation with a couple of friends. Goffman says that there is the potential to develop a sense of cohesion within a bounded group when there is a common focus and a common mood. We feel the solemnity of the ceremony; we follow along with the joke or the humorous incident being told by a friend and laugh at the punch line, or we give the appropriately surprised responses when being told that near friends are going to get married, or for that matter divorced. In the process, we also benefit from and contribute to the cohesion of the group. The sum of all of these small-scale rituals is the integration of our social network.

The mobile phone has become a conduit through which some of these processes take place. Clearly, a wedding carried out over the mobile phone would not have the same gravitas as a co-present event. However, we often use our mobile to contact our closest friends and family, for practical but also for expressive reasons. When we call to make arrangements, we also exchange funny or endearing remarks. We include slang and perhaps an emoticon in our text messages as a small bow to social dynamics. Because of this, the mobile phone is a medium through which we can easily perform these everyday rituals. Indeed the mobile phone gives us a direct and immediate channel of contact to exactly that person with whom we want to share the latest tittle-tattle or relate the latest yarn. Following from Licoppe's notion of connected presence, this idea of ritual interaction seeks to explain not only how we use the device to interact with one another, but also how that interaction becomes, in some small way, a part of the lore associated with the relationship. We call one another, and when in contact we engender a common mood and we coin a common sense of our shared experience. This in turn becomes a part of the ballast of the relationship. The mobile phone is one of the channels through which this occurs and, indeed, it is a particularly efficient one. Research shows that the mobile phone is clearly one of the most efficient tools for achieving integration (Ling, 2008).

Finally, towards the end of their edited volume, *Perpetual Contact*, James Katz and Mark Aakhus (Katz & Aakhus, 2002) propose a theory of "apparatgeist" to isolate and explore the core implications of personal communication technologies, of which mobiles are the most widespread and powerful. Apparatgeist describes a consistent, observable interplay between people and personal communication technologies, observable across contexts, cultures and life states. The "socio-logic" of "perpetual contact" (p. 308) is linked to the capabilities of the mobile telephone, but is manifested mainly in the set of motivations and orientations, both explicit and implicit, which people can be observed to hold towards the device. According to the authors, the promise of perpetual contact structures communicative behaviors in predictable ways over time, some of which can be seen in the ways people use the technologies to balance "competing needs for connection and autonomy" (p. 316), regardless of who they are or where they are. This balancing act, latent and overt, is much more central to one's actions with a personal/mobile communication device, relative to previous communication technologies and to face-to-face interactions – different enough, argue Katz and Aakhus, to merit a theory (explanation) of its own.

5 Theory and the road ahead – chapters 2–6

As is obvious, there are already a mosaic of perspectives, offering theories new and old to help illuminate the significance of mobile communication. Indeed, we have only presented examples from what is a rapidly growing and diverse literature. The next few chapters will move in and out of explicit mentions of theory. They will draw on works from other authors, occasionally placing insights from such radically different approaches as domestication and diffusion next to each other to make a point. This situation requires all of us to pick our poison – rigid reliance on existing theory risks obscuring what is truly new about mobile telephony. By contrast, new mobile-specific theories highlight what is new (absent presence, perpetual contact, etc.) but do so at the risk of underemphasizing continuity with previous technologies and with previous forms of mediated human interaction.

When discussing broader trends and the interrelationships between mobiles and society, we share more in spirit with the emergent schools of thought – those assuming complex and interconnected lines of causality – than with the determinists (for broader discussions, along these lines, see Fischer, 1992; Orlikowski & Iacono, 2001). Nonetheless, we are not social constructionists. The choices people make about how to use mobiles are not totally their own – handsets do some things well and other things not so well. Simple factors like the relative price of a call to an SMS, or what language the interface works in, have huge impacts on who uses these technologies, when and how.

It is helpful to distinguish between the identification of a trend or change (we see two – the boom in connectivity in the developing world and the arrival of ubiquitous reachability in the developed), and reflections on the implications for society of these trends. The latter task is our theoretical contribution. In that vein, we describe the emergence of a mobile logic. Drawing on others who have surveyed the same terrain, we too see a logic determined by (a) mobile and (b) increasingly ubiquitous devices. Mobiles now offer the promise of connection to all of us (not just the rich or lucky among us). From this, we see two implications. We see individuals apportioning their activities to a greater extent than previously possible; we see them interlacing multiple tasks with multiple actors and multiple venues, competing and jostling for time and attention. For society, or at least for groups of individuals, we see the implications of ubiquitous reachability playing out in the way individuals behave: not only that they are reachable, but that they expect others to be, as well.

This book will examine the various dimensions of mobile communication in everyday life. We begin with a brief overview of the history and current shape of global mobile telecommunications infrastructure in chapter 2. In chapters 3 and 4 we draw on a breadth of research to portray how the mobile phone is used in everyday situations. In chapter 3 we focus on first-time mobile phone owners in the developing world; chapter 4 examines the mobile phone use of five residents in the developed world. Rather than simply reviewing the research, we try to summarize different use situations as seen in the lives of fictional users. These range from small business owners in Africa, to teens in Japan and the USA, to poor farmers in India, to soccer moms in Scandinavia, to flashy business guys in Italy.

The vignettes are drawn from real situations and they have been reviewed and critiqued by others who have first-hand knowledge of the particular use situation. In some cases, they are patchworks of different individuals – others are made out of whole cloth. We adopt Weber's notion of ideal types in these vignettes (Weber, 1997). That is, we form the vignettes from elements that are to be found in real-life situations, but we do not intend to describe comprehensively all the characteristics of any particular case. Rather, we employ this device in order to illustrate some of the most common elements of mobile use. The point of these short fictions is to illustrate the variety of use situations and the way that the mobile phone has become engrained in our lives. In addition to the actual vignettes, we develop a commentary on the various dimensions of mobile phone use that are central to the use situations.

In chapter 5, we examine the way that mobile communication is a disruptive force in society. In this analysis we look at the rather innocent disruptions of mobile phones ringing in unexpected places as well as the more fundamental questions that the mobile phone poses to power relations and the structure of different institutions. We look at the way that mobile communication plays into the transgression of rules and antisocial activity. In addition, we look at the mobile phone vis-à-vis globalization.

Finally, in chapter 6, we examine how the mobile phone has introduced the notion of personal addressability and the interlacing of co-present and remote interactions. The ubiquity of the device means that we are perpetually available. These elements mean that there has been the development of a mobile logic. That is, the increasing spread of the device has meant that there is the expectation, and indeed the assumption, that we are potentially available to others, and in particular the others in our near social sphere whenever and wherever we happen to be. This mobile logic affects the way that we organize our daily lives, the way that we gather information and the way that we do our work. It is increasingly taken for granted, to the degree that we only see it when it is not there.

CHAPTER TWO

Short history of mobile communication

In the last chapter, we described the remarkable, rapid spread of mobile telephones (handsets and subscriptions) across the globe. In the span of just a few short decades, the technology has exploded, from being the glimmers in inventors' and hobbyists' eyes to becoming one of the most widely used consumer electronics on the planet, surpassing landline telephones and challenging television for ubiquity and desirability.

This chapter will describe some of the underpinnings, technical and institutional, which have made this spread possible. Although this might seem unnecessarily technical or commercial, it is an important precursor to the discussions about the social implications of mobile use found later in the book. It is tempting to treat the device as a complete, uncomplicated artifact (Fortunati, 2005a), as we tend to do with the everyday items which surround us. But the mobile, as it stands now, and as it will be in the years ahead, is the result of a myriad of experiments, technological breakthroughs, marketing experiments and policy interventions. Like that of the landline (Fischer, 1992) or the PC (Kidder, 1981) before it, the history of the mobile phone is a history of the interaction between people and technologies. Broader reviews of this history are available elsewhere (Agar, 2003; Levinson, 2004), but this short primer will set the stage for the chapters which follow.

It is important to note that the mobile telephony system is more than simply mobile handsets and subscriptions. Indeed the Nokia, the Motorola, or the NEC handset that we carry in our pocket is just the access point. When we place or receive a call and when we send or receive a text message we are engaging a whole system of equipment that encodes, transmits, switches, channels, notifies and decodes our shouts, whispers and emoticons. Without the towers, the switches, the exchanges, the protocols and the regulations supporting them, our fancy phones are just clunky MP3 players with bad cameras. It is the networks, not the handset, that allow us to connect (Cherry, 1977).[1] Indeed, as outlined by the Maitland Report, it is the economics of this massively complex system that allows (or prevents) the diffusion of telephony.

1 A primer on the mobile phone system

A confluence of factors made the modern mobile telephone possible: the development of radio-based communication, the rise of modern electronics (in particular the transistor), and the desire to provide expanded telephonic services beyond the landline network. The mobile telephone represents a marriage between radio-based communication (the kind used by police and fire fighters, and between ships) and the landline network of hard-wired telephones. To be sure, there are elements of both in the mobile telephone system. To understand the development of the mobile phone system, it is useful to look at how it identifies, tracks and facilitates individual calls.

In the first radio telephone systems for New York City, only a handful of users were able to simultaneously make a call. In that system there was only a single radio cell for a whole city. Advances in the way that the radio spectrum was used, and the development of the cellular concept – see figure 2.1 – meant that many more people could make calls using the same radio resources. These efficiencies came from the way that the mobile system dynamically apportions the radio spectrum to calls. In addition, the development of a mobile telephone system had to await the development of the cellular concept and the rise of the computing power needed to support the system. It was not possible to have a system that was able to keep track of where individual handsets were, which part of the spectrum was to be

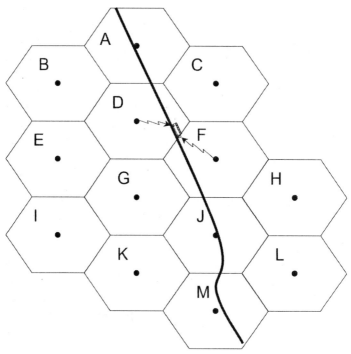

Figure 2.1 The layout of a cellular network with a highway crossing through the landscape

used on a call, the dynamics of handing off calls between cells and whether, for example, you were already making a call with the computing capacity available in the 1950s.

   When we carry a mobile handset with us, it is continually updating the system as to its location and status, so that the system can locate the handset should a call come in. As we, for example, drive from one side of the city to another, we pass through a series of smaller radio cell sites, each anchored by a radio tower.[2] It is from these radio cells that we derive the name "cellular telephony." While in an idealized version they form a series of contiguous hexagons, the exact size and extent of these cells depends on the local geography and the number of calls that the particular cell handles. When we receive a call, the system routes the call from the caller through the various exchanges to that particular radio cell where our handset happens to be and eventually to our own handset. This in itself is a minor miracle, namely that the mobile telephone system is able to keep track of and route calls between the thousands or the millions of individuals who are also inhabiting the same city.

Routing the call to the correct subscriber and to the correct radio cell is only part of the magic, however. If we happen to be riding in a bus when we get the call, in the eyes of the telephone system we are moving from one radio cell to another. Thus, the system needs to keep track of which radio cell will best transmit and receive the important message from our spouse that we need to buy milk at the store. As we move towards the edge of one radio cell, the signal between our handset and the radio mast becomes progressively weaker as we move out of its range. At the same time, the radio tower in the adjoining cell detects an active call in an adjacent cell is getting increasingly stronger. When the strength detected by the new cell reaches a certain threshold, the first cell hands the call over to the radio tower in the second cell. In figure 2.1, a bus is moving from cell D to cell F. Cell D is still handling the call, but Cell F is preparing to take it over.

All of this (usually) happens seamlessly, so that we are often not aware of it happening. Thus, we can concentrate on remembering the cheese and frozen pizza also on our spouse's shopping list, rather than on the technology. As is obvious, however, making or receiving a call via a mobile phone draws on a sophisticated and complex system of radio links, exchanges and computers for keeping track of individual handsets.

In all of this, it is also worthwhile to consider the complexity of the mobile phone handset. It is indeed one of the more intricate items we carry with us. Even in its most basic form, it includes a radio receiver as well as a sender. In addition, many mobile phone terminals also include a whole series of processors that often approach the sophistication of a personal computer in order to support other functions such as cameras, calendars, MP3 players, etc.

2 The development of cellular telephony systems

The mobile telephone system is obviously a legacy of traditional landline telephony.[3] It is the extension of a radio-based dimension, linked to the system of hard-wired telephones. Harnessing the radio spectrum for the purposes of communication dates back surprisingly far – to the 1860s – when Dr. Mahlon Loomis of Virginia in the USA was able to send and receive "electrical discharges" between mountaintops. These discharges could carry information, because senders could arrange them into the dots and dashes of the Morse code. Dr. Loomis received a patent for this work but did not pursue it commercially (Farley, 2005b). In the late 1880s the German Heinrich Hertz described how electromagnetic waves (what we call "radio waves") travel through the atmosphere. In the period that followed, several people started the work of employing this principle for the use of communication. The person who is most closely associated with its further development is Guglielmo Marconi (Agar, 2003). Marconi, who was half Italian and half Irish – his mother was an heiress to the McAllen whiskey family – started to send radio signals over greater and greater spans. Starting with the distance of 9 meters and moving progressively to 275 meters, 3 kilometers and the English Channel, finally, by 1901, he was able to send messages across the Atlantic Ocean. By 1910 Marconi was sending messages from the UK to Buenos Aires, a distance of just under 10,000 kilometers.

At the same time, the traditional landline telephone was becoming more common. The telephone had been around since 1876, with the critical addition of the telephone exchange[4] developed 2 years later. If we go another 25 years, the number of telephones was about 1 subscription per 10,000 people in 1900 in the USA. By 1915 this had risen tenfold to approximately 1 per 1,000 people. Thus, while still a relative rarity, the telephone was well established by the beginning of the twentieth century (Fischer, 1992; Pierce, 1977).

Moving back to the development of radio, the first practical application for this form of communication was shipping. By the first decade of the 1900s, many ships were outfitted with Marconi wireless devices. The event that assured the place of wireless communication in shipping was the Titanic disaster (Goggin, 2006). While the Titanic was sinking, it sent out emergency messages via its radio system. Tragically, the radio operators on nearby ships had signed off for the evening and thus did not receive alerts about the situation. This episode and its grim results underscored the potential of wireless communication. In the wake of that incident, ships were required to have 24-hour radio availability.

In 1906, Lee De Forest invented the vacuum tube. This advance made the transmission process more energy-efficient and it allowed for the practical transmission of voice (not just Morse code). In addition to enabling two-way radio communication, this also opened the way for the development of broadcast radio. This was used in a variety of situations, including commercial radio stations (Agar, 2003; Lindmark, 2002), by police[5] and fire fighters, in the dispatching of taxis and, for example in New York Harbor, the dispatching of tug boats. Radio was also used in communication with ships and airplanes. In addition, the military adopted radio communication. Thus, by 1920, we have the two basic elements for the development of mobile telephony. There was some speculation as to the ability to combine the two media (Brooks, 1976), but this development would have to wait until after the Second World War and the arrival of the computing power needed for the management of the system.

After the Second World War, the work on the transistor resulted in the development of smaller and more energy-efficient communication devices. The first commercially available mobile communication systems were developed in this period. In 1946 the first radio-telephone service was established in St. Louis, Missouri. This system was basically the same type as that used for police dispatch, linked to the traditional telephone system. It did not employ the cellular approach that was later adopted. Thus, the system was limited in its reach and limited in the number of subscribers who could use it. The devices used for calling were large – they were almost always mounted in a car – and their use demanded a lot of energy (Lindmark, 2002). Farley notes, for example, that, when using a car-mounted Swedish system, the headlights of the car would noticeably dim when the user made their call (Farley, 2005a, 23_4).

The idea of having a large number of smaller "cells," each with their own transmitter and with the ability to hand off calls as the individual moved from one cell to another, arose in the late 1940s at Bell Labs in the USA (Lindmark, 2002). Thus, instead of one central transmission site that covered a large geographical area such as a city, the system called for dozens of smaller transmission towers that would keep track of the location of the individual handsets in case a call was to be routed to them, and would also deal with the handing over of calls that were in progress.[6] This, along with the development in 1948 of the transistor – and the subsequent addition of the integrated chip – was central in the growth of the mobile telephone system. Indeed, the sophisticated computing power required to track and handle mobile handsets had to await the development of computing power that was not readily available in the 1940s and 1950s (Agar, 2003).

In 1969 the cellular system was introduced as a pay phone on the Amtrak Metroliner between New York City and Washington DC. Four years later, Martin Cooper used the first hand-held mobile phone, which had been developed by his group at Motorola. The technology was starting to be in place, but there were still several important issues that needed to be sorted out before mobile telephony could take off. These included regulatory issues, sorting out the welter of different standards, the refinement of the handsets and finally the commercial development of subscriptions that would appeal to users.

Mobile communication handed regulators a new issue. On the one hand, they faced the need to apportion the radio spectrum. This limited resource has to be shared by a wide variety of users. The military, commercial broadcasters, private hobbyists and commercial interests of various types all seek to have portions of the radio spectrum. If different users try to transmit and receive information in radio frequencies that are too close to each other, the two signals will interfere with each other and neither will be of any use. Thus, the regulatory bodies in each country – and also internationally – must agree as to who can transmit on what frequencies. The purchase and sale of these rights is big business.[7]

The history of the device shows that there was also a significant role played by regulatory authorities in the different countries. While, for example, there were extended interactions between industry and governmental authorities in the USA (read: delays), there was a push for transnational standardization in Europe. The model in Japan saw extensive cooperation between government and the different industrial actors (the phone companies, equipment producers and the handset producers).

At the dawn of the mobile phone era many countries simply apportioned the frequencies to the national telephone operator. In the USA, however, the Federal Communications Commission (FCC) – in the wake of the dismantling of the former AT&T monopoly – used a market model to develop mobile communication (King & West, 2002). They divided the country into approximately 300 metropolitan markets and 428 rural service areas.[8] In almost all of these areas, the FCC granted two licenses. There was relatively careful analysis of applicants for the 30 largest markets but a lottery system of selection was used for the progressively smaller markets. According to West (2000), there were 96,000 applications for licenses in 215 of the less central markets (see also Agar, 2003). This approach brought a wide variety of nascent operators onto the scene, who bore with them a wide range of competence and a wide variety of technologies. Mobile communication in the USA during this period was characterized by many small operators with poor interconnection between them, which presented difficulties for users when roaming outside their core area. All of this stifled mobile communication in the USA for several years.

At about the same time in Europe there were still nation-wide operators – albeit on a smaller scale – who were able to develop one mobile phone system for the whole country. In the early 1980s the Europeans, in the spirit of collective good, eventually came together to develop a common standard – the Global System for Mobile Communication (GSM) – that could be used across the whole continent. The development and commercialization of the GSM standard resulted in a rush of customers. TeleDanmark, for example, planned for 15,000 new GSM customers in 1993, the year of its commercialization, but acquired 65,000, and Sonofon in Finland planned for 25,000 in 1995 but acquired four times as many (Haddon, 1997). The GSM has gone on to become the dominant standard for mobile communication around the world.[9] In many countries, including the USA, an alternative standard named CDMA is also used. Indeed there are dual-mode handsets on the market, which allow users to switch between GSM and CDMA networks, depending on availability and user needs.

Although GSM and CDMA are the dominant standards in many locations, there are other variations. The low-cost "little smart" system in China is neither a landline nor a standard cellular telephone system, but rather a kind of "amplified cordless phone"; its 50 million users can use the phone as they move about within their cities, but cannot roam much beyond them (Castells et al., 2007; Qiu, 2007).

Around the world, different countries have taken different approaches to the sale and management of the mobile spectrum. Developing countries, in particular, have had to balance the demand from customers for low rates with the opportunity to sell licenses to raise revenue, and with the need to find the right mix of competition vis-à-vis fixed-line "incumbents." Many of the latter have traditionally been sources of revenue for governments; they can be state-run, the state can be a major investor, or the state can simply enjoy steady revenues from taxing calls, particularly international calls (Beardsley, von Morgenstern, Enriquez & Kipping, 2002; Best, 2003; Ibarguen, 2003; McDowell & Lee, 2003; Pereira-Filho, 2003). Indeed, what is best for spreading mobiles quickly in urban areas might not be what is best for rural areas, which have tended to be less cost-effective to service and therefore less attractive to mobile operators. Some governments have made rural service part of the license requirements for mobile operators. Still others are working to design technical or policy interventions to bring service to rural areas which operators might otherwise choose to avoid (Dymond & Oestmann, 2003; Engvall & Hesselmark, 2004; Galperin & Girard, 2005; Hudson, 2006).

3 Data options

Of course, it is not just about the voice call. Although sales of "airtime" or "minutes" still make up the bulk of most operators' revenues, the networks can carry data as well (PT, 2008). This has taken many forms, from the humble SMS to the super-fast 3G internet connection.

Another development that came with the GSM standard was texting – the so-called Short Message System (SMS) (Trosby, 2004). For the first years of its existence, it lived a quiet life without much attention. In about 1997–8, teens discovered the potential that was, at that time, free to use. This discovery spawned a new form of mediation that in turn resulted in new forms of interaction and new linguistic formulations (Baron, 2008; Hård af Segerstad, 2005; Ling, 2005b, 2007b). By 2005 it was estimated that a trillion text messages were sent annually (Ling, 2008) and it is estimated that by the time this volume is published it will be 2 trillion per year (Research and Markets, 2008). Texting has proven to be a remarkable win–win for customers and carriers. Customers love the per-unit pricing and flexibility of the simple format; operators love that, when measured by revenue per byte of information, text messaging is remarkably lucrative. As sites like 160characters.org attest, and as we will discuss later in the book, the platform has spawned junk mail, flirting, chat rooms, banking and almost everything else under the social and economic sun.

In addition to voice and text communication, there are a variety of other services available via the mobile phone. These include I-mode and the similar Wireless Application Protocol (WAP), Multi-Media Messaging (MMS), various types of mobile internet access, and services such as mobile TV. WAP is more of an open standard that functions somewhat similarly to the internet's HTML.

A successful service in Japan is I-mode (Lindmark, 2002). Strictly speaking, I-mode is not a standard in the same way as GSM and CDMA. Rather it is a product sold by the operator DoCoMo that, along with voice telephony and texting, gives one access to a "lite" version of the internet and to a variety of services. The user of I-mode can download various types of information (weather, news, horoscopes, etc.). I-mode is a so-called "walled garden" where there are DoCoMo-certified services that the user can access. The content owners and DoCoMo have a revenue-sharing plan in which a certain portion of the money generated by a service goes to each. The idea of Apple iPhone Apps is somewhat similar. These services occupy a "walled garden" in that they are authorized by Apple (or DoCoMo), and the billing occurs via the normal telephone bill or, in the case of the iPhone, via the user's credit card.

WAP is not nearly as popular as I-mode. Technical issues and inflated expectations and lack of willingness to pay have resulted in the notion that WAP was a failure. The mobile internet has eclipsed WAP. It uses the same protocols that are used via the PC. While it is slower (and more expensive) than the broadband available in homes and offices, mobile internet is being used on advanced mobile phones for a wide array of services. This development is being facilitated by mobile phones that can send and receive data at higher rates. There has been a progression of services that facilitate faster data transmission. These include systems such as General Packet Radio Service (GPRS), Enhanced Data rates for GSM Evolution (EDGE) and High-Speed Downlink Packet Access (HSDPA). While the speed of these radio links is much faster than users experienced with the earliest mobile phones, radio technology is unlikely to reach the transmission levels of cable-based services such as the various forms of Digital Subscriber Lines (DSL) and fiber optics. I-mode and WAP are both mobile internet protocols. I-mode is a proprietary service owned by the Japanese mobile operator DoCoMo.[10] It has proven to be popular in Japan and has been exported to several other countries, though with less success.

Another service that was seen as a legacy of text messages was Multi-Media Messaging (MMS). This is an enhanced version of messaging where the user can attach photos and sound in addition to text. MMS is living a comfortable, but somewhat quiet, life as an additional service for mobile phones. In some cases camera phones are being used to take photos and to send them via the mobile telephone net. A more common practice it seems is the use of the camera phone to take a picture. From there it is transferred to a PC via a cable or Bluetooth and then spread to the appropriate friends and family via sites such as Facebook or MySpace.

There are other services that are being developed. These include things such as mobile TV, mobile payment systems that use Near Field Communications, location-specific services that are based on Global Positioning Systems, using the mobile phone as an MP3 player, etc. Some of these services will become commercially viable while others may not reach that status. Thus, between 1969, when the first cellular system was put into use on the Amtrak Metroliner, and the end of the century, mobile telephony moved from being a rarity to being a well-integrated portion of our everyday lives.

4 Mobile handsets

Along with the development of the cellular system, there has also been a parallel development in the handsets that we use to make and receive calls. They have gone from being large devices that were basically intended to be mounted in cars – and that seriously drained the car's battery – to the sleek small devices we now carry in our pockets or our purses. The early mobile telephone handsets[11] were mobile only in the sense that they needed a motorized conveyance to move them about. An early Nokia, for example, weighed 9.5 kilos (21 pounds). However, the development of transistors and eventually integrated circuits has reduced their size and heft. By the time that Martin Cooper made his fabled call on the first handheld device (a Motorola Dynatac), it weighed in at just under a kilo (2 pounds) and was about the size of a large package of spaghetti. It had a battery that would give the user a modest 30 minutes of talking time.

Several elements played into the development of smaller and smaller handsets. As scales of production came into play, manufacturers were able to move away from the larger "brick" phones towards handsets that are not out of the scale that we use today. This involved the integration of more and more functions onto the same chips. This in turn meant that the devices required less battery capacity. In addition, the growth of the mobile telephone system meant that the density of cellular sites increased. This tighter grid of sites meant that the handsets needed to use less power to communicate with the cell tower, reducing the need for huge batteries.[12] This can be seen, for example, in the development of the Motorola MicroTAC (11 oz) and, further, the StarTAC that weighed only 90 grams (3.1 oz) (Goggin, 2006; Steinbock, 2003). Most mobile phones now weigh between 100 and 200 grams (3.5 – 7 oz).

The transition to digital systems in the early 1990s – most notably GSM in Europe – facilitated the development and spread of handsets (Bastiansen, 2006). A new dimension with GSM was the addition of a separate Subscriber Information Management

(SIM)[13] card (Goggin, 2006). The SIM card allows the division of the operator and the handset manufacturer. Thus, the owner of a Nokia handset may subscribe to T-Mobile, Telenor or Vodaphone. The subscription is separate from the handset and can be exchanged as needed. This has given rise to SIM switching (Wong, 2007) – a practice particularly common among cash-constrained users in developing countries – in which the individual may change the SIM card depending on where the best connection or the cheapest rates can be found.

The mobile phone has become a Swiss Army knife-like device. We carry it in our pocket and it has morphed into a personal communications manager, with the potential to "eat everything." It comes in a variety of forms, colors and sizes, and with a spectrum of functions. It is a camera and a photo album with a high-resolution color screen and it is an FM radio / MP3 player / video player. It is a gaming terminal and it is an internet access point where we can read e-mail, IM and access social network sites. It is a calendar and a contact manager. It is a GPS navigation system. It allows for text production and the development of presentations and other office-related functions. The mobile phone can also be a payment device not unlike a credit card. Indeed, the mobile phone can do the jobs of many of the items we currently carry in our wallets or purses (aside from, for example, water bottles and the like). In the future we may not need separate bus tickets, credit cards, pictures of our sweeties, music playing device, reading material, etc. All this will be included in what started as a phone and a texting gadget.

Mobile phone handsets often live several different lives. Some start in the developed world and then are refurbished and sent off to a second life in other parts of the world. There is, for example, a lively trade in used / refurbished / grey- and black-market mobile phones in, for example, Hong Kong. Anthropologist Gordon Mathews suggests that as many as 20 percent of the mobile phones in sub-Saharan Africa are bought and sold in the Chungking Mansions in Hong Kong, by merchants who travel between there and Africa literally carrying suitcases full of mobile phones (Mathews, personal communication, 2008). Jan Chipchase, a researcher at Nokia who studies mobile phone use in a variety of settings, describes finding a broad variety of reconstituted phones. He recounts finding, during one trip to China, Nokia phones retrofitted with batteries from other phones sporting Malaysian software and the contacts of previous owners still in the memory. These could be bought for 22 euros (approximately $20).[14] It may be that as many as 20 million refurbished handsets are being bought and sold every year. As many as 60 percent of all phones sold could be recycled devices (ABI Research, 2007).

Finally, those phones that are too decrepit to be retrofitted, resold or cannibalized contribute to the growth of so-called "cyber junk." In 2007 about 780 million handsets were retired worldwide (Basel Action Network, 2004). While mobile phones are much smaller and consequently less of a burden than, for example, PC displays, they contain a variety of toxic wastes nonetheless. The batteries, circuit boards, displays and the housing of the phones contain various chemicals and elements that are toxic when not treated properly. The materials include lead, beryllium, arsenic and cadmium. Interestingly, they also include silver, copper, platinum and gold (Mooallem, 2008), worth as much as a dollar if re-claimed from the scrap. If not treated properly, these wastes could enter into the ecosystem and become a problem. The proper disposal of mobile phones can address much of this issue, but alas, the proper thing and the economically possible thing are not always the same. Serious actors in this area take phones, separate out the usable from the non-usable ones and then render the non-usable phones into their component parts in an environmentally benign way.

5 Business models, commercial issues and their impact on adoption around the world

In addition to the technical and the policy issues, the development of mobile communication demanded new approaches to commercialization. These currently range from prepaid "top-up" subscriptions to "family plans" to the more traditional postpaid subscriptions.

Prepaid subscriptions are perhaps one of the keys to understanding the broad adoption of mobile communication. Prepayment for access to landline telephony has long been used for people with poor credit.[15] We also "prepay" when, for example, using a traditional landline telephone booth. Bringing this into the era of mobile telephony, prepaid subscriptions are ideal for persons with low use and for people with irregular income (teens and often lower-income people in developing countries). Kalba writes:

> In retrospect, this introduction of prepaid technology, considered a peripheral achievement at the time, may have been the most significant innovation since the development of the cellular concept and its initial implementation. Without prepaid, which consists largely of storage and billing software, mobile calling may not have reached as many as two-thirds of today's subscribers, especially those located in poor and moderate-income developing countries, where participation in the cash economy is often a stochastic or itinerant activity. (Kalba, 2008, 642)

The fact that there is no monthly fee and that you only pay when you need to "top-up" your account means that it is ideal for people in these situations. The disadvantage from the users' perspective is that, since there is no fast monthly fee, the operators often charge more "per minute" for prepaid subscriptions than the per-minute price for postpaid subscriptions.

In some cases, the sale of prepaid "top-ups" can be in extremely small denominations. While in Scandinavia the smallest prepayment might be approximately $25–30, in developing countries it is possible to prepay as little as $0.50. Those with irregular income or only the occasional need to call can use this system in order to secure access to the mobile telephone system. Indeed, for some there is not even the need to own a handset. Rather, they simply have a SIM card (containing their subscription information and subscription credit balance) that they insert into a borrowed handset. In some countries, literally all subscriptions are prepaid. According to Kalba, 100% of those in Nigeria are prepaid. Prepaid subscriptions encompass 99% of all subscriptions in Iraq and Pakistan, 98% of those in the Philippines, 97% in Algeria, 96% in Bangladesh and so on. By contrast 3% of users in Japan and Korea, 5% of Finnish users, and 14% of those in the USA have prepaid subscriptions (Kalba, 2008). For the world as a whole, 46% of all subscriptions are prepaid (Gray, 2005). Prepaid subscriptions are also common for teens and youth in, for example, Norway.

Another dimension of mobile telephone subscriptions is how the charges are divided between the interlocutors. In most countries and with most operators, a system known as "calling party pays" is used. As the name states, the person who initiates the call also pays for it.[16] The system that is common in the USA uses another approach. There, the individual subscriber often pays a monthly fee in exchange for "airtime" minutes (for example, $50 for 1,000 minutes). In daily use, whenever a person is talking to another person on their mobile phone, this time counts against their minutes regardless of whether they make or receive the call.[17] While in practice people may not pay this much heed, it actually means that when calling another person you are implicitly asking them to spend some money on you. It is sort of like inviting yourself to dinner. In some cases, this has meant that people turned off their mobile phones to avoid receiving calls/ charges. They would only use their phone to call out. In most ways, this defeats the purpose of having a mobile phone in the sense that it reduces its utility in the broader social sense.

Finally, mobile telephone handsets are also a part of the general subscription landscape. On the one hand, there is "lock in" for some popular mobile phone handsets and there are various forms of subsidies associated with the sale of subscriptions/ handsets. Looking at the first of these, mobile telephone operators such as AT&T, Vodaphone, DoCoMo and T-Mobile often want to be the only one offering the most popular handsets. If a particular operator is the only one offering such a handset, then they will be able to retain existing customers and recruit new ones. For their part, handset producers (Nokia, Motorola, Sony Ericsson and Apple) want to sell as many of their popular handsets at as large a margin as possible. If the handset producers and the operators are able to agree on conditions assuring that the producers will receive adequate promotion and that the sales margins will be ensured, they will then license a particular operator as the exclusive agent for a particular handset for some time period.[18] The alternative is to sell the handset model through as many outlets as possible.

From the perspective of the customer, when they purchase a subscription and a telephone handset, the operator often subsidizes the handset. In exchange, the operator 'locks' the handset, making it functional only in the operator's network. If the customer is interested in having an "open" handset – that is, a handset that can be used on a different operator's network – they will have to pay a much higher price – indeed, this can be as much as 10 times more. The point is, of course that the operator wants to keep the user "locked in" to their network as long as possible. It is possible, albeit a violation of the contract with the operator, to "hack" a locked phone so that it can be used on different networks.

In addition to the pricing of handsets for new customers, there are different ways of trying to reduce what they call "churn" (subscribers leaving a particular operator's network) by subsidizing their telephone handsets. When the contract period expires, there are various forms of bonus points or offers for new handsets designed to entice the user into renewing his or her contract period with the same operator. With unlocked GSM telephones, the user is, as noted, free to have several different subscriptions, as long as he or she can secure and swap between different SIM cards. This allows them to use the least expensive, or the most accessible, subscription, as needed.

Bringing all this together, from their rather unwieldy beginning, mobile phones have captured our imagination with their ever more sleek and exciting products. There are well-designed Nokias, elegant Motorolas, Sony Ericssons with cameras and music players, HTCs for the business guys and gals, alluring Apple iPhones, plus dozens more. Different players arrive on the scene and others make their more or less planned exits. All these contain materials that come from around the world and that again have a global manufacturing pedigree (Agar, 2003; Ling, 2007b; Ling & Baron, 2007). As noted above, they can be a $15 bootlegged device or a $75,000 platinum and diamond version – each being a significant status symbol in their respective contexts. Mobile phones have become an icon of our age. They have become a basic part of everyday life and they will be one of the symbols for which we are remembered.

CHAPTER THREE

Mobile communication in everyday life: 3 billion new telephones

1 Introduction

As we discussed in the introductory chapter, some mobiles are complements to landlines – they are owned and used by people who also happen to have landlines. In the developing world, however, the mobile is often a *substitute* for the landline (Hamilton, 2003). Thus, it brings many of the same fundamental benefits (and social and economic changes) associated with landline communication (de Sola Pool, 1977; Fischer, 1992) to hundreds of millions of people who previously could not afford or did not have access to a traditional landline.

In that spirit, we now turn to five vignettes – representations of different everyday uses of the mobile phone – of five people whose first and only telephone is a mobile telephone. In chapter 4, we will present another five vignettes of people who have had access to both landline and mobile telephones. The first two describe actual people, Rohit and Annette. The characters in the rest of the vignettes are composites; the behaviors and attitudes are aggregations of behaviors of people we have observed in our own fieldwork, and have assembled from researchers from around the world. Though drawn from primary research, the resulting portrayals are fictional illustrations, designed to introduce key themes parsimoniously.

1.1 Rohit, Bangalore, India

Rohit, 25, works in a tech support center in Bangalore, India. He usually lives by himself. The vignette is based on an interview appearing in C. Y. Wei (2007, 202–3), augmented with further conversations with Wei on October 19, 2007.

Phone type: feature phone, no GRPS/WAP
Subscription type: prepaid "top-up" subscription
Average monthly expenditure: $15
Hours of work to buy his mobile phone handset: three days of work to buy a 4,000-rupee ($100) handset
Most common use: voice calls, texting, games, FM radio and alarm clock

Rohit came to work at his current job in the technical support center of a major software firm, leaving his hometown of Delhi, over 1,000 miles away from Bangalore in the north. His parents keep in close touch with him, and in fact, his mother has recently stayed with him for a long visit. He works an overnight shift. Prasad, a friend from home, is staying with him while he looks for a job in Bangalore. When Rohit comes home early in the morning, his friend is getting up to start his day.

Rohit has two mobile phones, each with its own subscription, which he regularly uses: his main Bangalore phone as well as one with a Delhi number. The Delhi number is paid for by his parents because they have a plan that allows them to call him for free and without roaming costs. This plan helps his mother stay in close contact with Rohit – her only son, her pride and joy. Rohit normally carries only the Bangalore phone. He reserves the Delhi phone for family calls and a few friends at home to save on the expense for his parents.

Rohit works in front of a PC all day, but still uses the mobile to stay in touch with friends and family, even during his work shifts. Often, he is just planning where he will meet his friends for coffee or a movie, but he does so despite signs in the office which remind the employees that they are in a professional setting and need to keep mobile chats (and ringtones) as quiet as possible. Not surprisingly, he texts a lot, sending at least 10 SMS messages a day.

His apartment is pretty nice for a guy of his age – he has a fair bit of furniture and lots of family photos; he also has a home PC. Besides the two main phones, there are two other mobile handsets on his kitchen counter. He keeps one of these handsets handy but does not use it. The other handset he keeps in a basket – a broken Nokia that he had smashed in a fit of rage after a fight on the phone with his on-again, off-again girlfriend who lives in Delhi. The basket also holds a collection of chargers. These four phones were all purchased in the last five months. Rohit said that if we had met him earlier, he could have displayed even more handsets from when he switched from GSM to his current CDMA telephone.

Rohit has been in his on-again, off-again relationship for five years. They are now "off" – she is engaged to another man. This outcome, according to Rohit, suited his parents and his friends who disliked how the two of them fought. After the breakup, Rohit chatted over the computer with a friend about the latest quarrel he had with the girl, and the friend urged him to cancel her mobile subscription (which he was paying for). His friend told him to cancel it right now, "in front of him." With the support of his friend, Rohit immediately called the mobile company to cancel.

1.2 Annette, Kigali, Rwanda

Annette, a 34-year-old woman in Kigali, is a mother of one, and a restaurant owner. An earlier version of this vignette appears in Donner (2005).

Phone type: Nokia color display, with camera
Subscription type: prepaid "top-up" subscription
Average monthly expenditure: $15
Hours of work to buy her mobile phone handset: about a week's worth of her restaurant's profits, to buy a $75 used handset
Most common use: voice calls, texting and intentional missed calls

Around lunchtime, Annette has to divide her attention between preparing the sauce for her Matoke (steamed plantains) and fielding calls from her regular customers. Since 1996, Annette has run a small restaurant in Kigali, Rwanda, serving mostly Ugandan specialities to the many Ugandans living in the city. She's on her third handset now, a lot "fancier" than the one she started with in 2002, and, indeed, even fancier than the nice Sony she bought to replace it. Some petty thieves got to that Sony early last year. Like almost everyone she knows, Annette uses pay-as-you-go "recharge" cards to top-up her account. She pays only for the calls and texts she makes, gets free incoming calls and texts, and trades beeps (free, intentional missed calls, with coded meanings),[1] as long as she keeps a minimum balance active on her account.

Some of Annette's calls are with friends and family, particularly her mother. The local police recovered her stolen handset a few weeks after she had lost it. She had already replaced it, so she decided to give the extra handset to her mother; they now chat from time to time. She also calls her daughter, who is living at a school back in Uganda, hundreds of miles away. Needless to say, the mobile helps Annette feel closer to her family.

But she has found that having a mobile makes her daily business transactions go more smoothly. The restaurant has never had a landline, let alone a PC, but now with the mobile phone Annette is reachable all the time. Her customers have her number. Some call in advance to order lunch. She says, "The food is always on time and easier for them. Not like first reaching here and ordering. No sooner do they park than we put lunch on the dining table, since we are aware of what they will have. So they can use only ten minutes to eat." Stragglers might instead send a "beep" to check if she still has food left for the day. If she does, Annette will "beep" back, hanging up on her customer on purpose. By hanging up on each other in this way, neither Annette nor her customer had to pay for a call or even a text message, but both of them know the code that works for them at lunchtime . . . in this case, a pair of beeps means "come on down." Matoke takes a while to cook, but, thanks to her mobile, Annette is now in the fast-food business!

1.3 Prashant, Warana, India

Prashant, 31, is a sugarcane farmer in Warana, Maharashtra, India. His extended family lives with him. The vignette is based on Veeraraghavan, Yasodhar & Toyama (2007), augmented with an interview with Veeraraghavan, on October 20, 2007.

> **Phone type:** various shared handsets
> **Subscription type:** pre-paid "top-up" subscription
> **Average monthly expenditure:** $1
> **Most common use:** voice calls, checking inventory and balances via SMS

Like his father, his brother and most of his friends and neighbors, Prashant is a sugarcane farmer, in the village of Warana, about 400 kilometers from Mumbai. His family, including his wife, his parents, three children and two brothers (and their families), live with him.

The sugarcane business is hard work, but his family is better off than those who must eke out a living, only eating food they grow themselves. Instead, he belongs to a cooperative – 50,000 farmers strong – all working together to sell the sugarcane harvest to the domestic market three times a year. In years when the rains cooperate, Prashant and his family earn about 60,000 rupees, or $1,500, for their work. Everybody, including his wife, the kids and his brothers, helps out. Odd jobs and informal enterprises, like milking cows (to sell the milk) or a bit of carpentry, help earn the family another 15,000 rupees ($375) a year.

Prashant had seen the ads for mobiles on TV (though he can read and write, he does not have time for newspapers or magazines), but did not see one up close until last year, when the first mobile tower was built near his village. He has power on his farm, but goes out frequently, especially during the rains.

Soon, he noticed some of the wholesalers and truckers passing through had mobiles, and the wealthier farmers in the village soon followed.

The first time he used a mobile, it was alongside the information agent who worked for the sugarcane collective. It was a remarkable, surprising display to all the farmers present that day. The information agent asked for Prashant's account code, and, seconds later, he could see details from Prashant's accounts with the cooperative, right on the handset's tiny screen.

For the previous decade or so, a portion of Prashant's dues to the cooperative had been paying for some electronic information kiosks and agents to run them. These kiosks would use some rather technical protocols on an older PC and a dial-up network connection to pull down price information about sugarcane, and would also track the balance of each farmer – credits owed from sugarcane sales, and debits owed due to use of water and fertilizer, etc. The kiosks themselves were a great help, since farmers only had to wait a couple of days to get their balances and to learn how much a harvest had yielded for their families, compared to two or more weeks when using the traditional paper system. But the network of village kiosks was expensive for the cooperative to maintain, the internet connection was flaky at best, and power cuts often closed the kiosk in the village.

In 2006 the collective tried an experiment, replacing many of their kiosks (including the one in Prashant's village) with a mobile phone-based system. Using this system, the former kiosk operator still accessed the master PC at the cooperative's headquarters, holding the cooperative's overall records. However, unlike with the old system, the kiosk operator could now carry the cooperative-owned mobile and roam from farm to farm, even visiting the farmers while they were still in the fields. All the same account information is available, but the new mobile system is faster to update and cheaper to maintain and install. It is also much more timely and convenient for the farmers to access.

Now that he can check his balances more frequently, Prashant feels more in control. Those few days before he learns how much revenue his crops have earned are often emotionally stressful ones. Thus, it is nice to get the information as soon as possible. This allows him to plan the next planting. He can make better choices about when to plant, and how much water and fertilizer to use. In addition, it allows him to catch mistakes. Last year the person weighing Prashant's sugarcane inadvertently entered the wrong number. Since Prashant had the actual receipt, he could quickly check the information and correct the mistake. The ability to catch this problem quickly saved him and his family a lot of trouble.

Lately, Prashant has been visiting his friend Mohan, another farmer down the road. Mohan has bought a used handset. On occasion Prashant can borrow the handset to stay up-to-date even on days when the information agent does not visit with the collective's handset. Prashant and Mohan have agreed (without really having to say anything) that, as long as it is just an occasional thing, Prashant can send the SMS to the cooperative's agricultural information system without paying Mohan directly. If, however, Prashant wants to talk to a relative or perhaps order some supplies from the city, he will come with a few rupees in his pocket to share with Mohan's family. Prashant has found that using Mohan's phone is better than waiting in line at the public payphone (PCO) in Warana. In addition, it is more private, since when he talks on the phone in the village everyone in town can hear his business.

If he gets another good harvest, he is considering buying a handset of his own. Like Mohan, he will go to Kolhapur, a town about 10 kilometers from his village, and will buy a second-hand one, with a black and white screen, from a vendor in the market. He hopes he can get one for 600 rupees (about $15). He has his eyes on the Nokia 1100 – he likes the flashlight, and the way the handset has a keypad in his language, Marathi. He also hears the long-life battery is great.

1.4 Amihan, Singapore

Amihan, 37, is a Filipina domestic worker living in Singapore. She lives in the home of her employer. The vignette is based on research by Ellwood-Clayton (2003, 2005), Paragas (2005), Pertierra (2006), Sun (2006), Thompson (2005), Uy-Tioco (2007).

> **Phone type:** Nokia 1600 (a simple color phone)
>
> **Subscription type:** prepaid "top-up," with a Mobile Network Virtual Operator (MNVO) which matches the brand and services of her family's mobile carrier back in the Philippines
>
> **Average monthly expenditure:** $30 in text and voice calls, $20 in shared load, transferred to her daughter
>
> **Hours of work to buy her mobile phone handset:** 40
>
> **Most common use:** voice calls, text messages, money transfers

Amihan can still remember the first time she made a mobile call from her room in Singapore. She called her daughter, Dalisay, back home in Roxas, in the Philippines. Ever since, it has made the 2,000 kilometers between them feel like just a few meters, if only for the few minutes they are on the line together. Amihan has been a housekeeper in Singapore for seven years, travelling back to visit her daughter and parents only once a year. It's been hard being away so much, but, like many Filipinas, she decided that being an overseas foreign worker (OFW) was the best way to provide for her family.

Her mobile has quickly become the most important way she stays involved with the goings-on at home. She is happy that her parents, who are getting on in years, can contact her anytime they need to, and that she can call them to remind them to do important things around their house. Dalisay is almost 16, but sometimes it seems she is not comfortable sharing what she is thinking or feeling with her mom. So, Amihan has learned that text messages – little notes about important things, and not-so-important things – are a better way to keep the bond between them. Lately, Dalisay has been forwarding some SMS messages from the "text God" service back in the Philippines. Amihan is not sure she believes the messages are really from God (or even blessed by him), but it is still a way to encourage her family's faith. It is better than the other SMSs she sometimes talks about with her daughter. Dalisay has a bunch of "male admirers" – some she barely knows – who text her at all hours of the day. Dalisay assures Amihan that she's not "with" any of these textmates, that they are just friends and sometimes flirt a little, but, of course, as her mom, Amihan is worried. The fact that Dalisay says that all her girlfriends her age also have these textmates does not really make Amihan feel any better.

On a happier note, a great thing about the mobile is that it has made it easier for Amihan to send money home (that's the whole point of working abroad, right?). She uses her mobile to send home about 7,000 Philippine pesos (about $175) a month. Before she bought the mobile, she would send money via a remittance company (which took almost 5 percent of her wages as a commission), she'd send money by post (which took a long time to reach her family) or she'd just hang onto the money until she made the trip home. Now, she just goes to the local office of her mobile company and hands over some of her wages. For a low fixed fee (a couple of dollars), that money is "sent" immediately to her daughter's or her mother's mobile phone. They get a text message and a code saying the money has arrived, and take it to the store in their *barangay* (village) to get the money. Until her family has a bank account (which might be a while), this is the easiest way she can imagine to send money home.

There is really only one person who is unhappy that Amihan has a mobile, and that is her long-time employer in the house in which she lives and works. Her boss complains that Amihan is always chatting with her friends (other Filipina housekeepers) in Tagalog – a language her boss will never be able to understand. Amihan thinks her boss is just nervous and feels like she can't control Amihan's every little move (the way she used to), and Amihan likes it that way. She may be stuck in a tough job, but at least it gives her the power to take a break when she wants, and to get back to the *barangay* for a few minutes, if only in spirit.

1.5 Liang, Guangxi, China

Liang, a migrant worker at a chemical plant in Wuzhou, Guangxi, China, is 26 years old. He lives in a company dormitory. The vignette is based on research by Cartier, Castells & Qiu (2005), Gordon (2007), Law & Peng (2007), Su (2005), and Yu (2004).

Phone type: used Nokia 8850
Subscription type: prepaid "top-up" subscription
Average monthly expenditure: $10 voice and text, plus $20 on the "party line"
Hours of work to buy his mobile phone handset: 250 yuan phone ($36), about 4 days' work income
Most common use: voice, text, chat room

Wuzhou is not the biggest city in Southern China, but it still held lots of promise for Liang when he moved there from his village in Yunnan province. When he first made the move, he was looking for a better job. His dad passed away and he really did not have anyone left in his home village to stay with. Because of this, he just took buses for a few days and reached Wuzhou. After a couple of false starts, he landed a good job with a growing company. He lives in a factory dormitory at the chemical plant. He has been a good worker and has gained a lot of skills since he started, but the company has a lot of power over his life. He did not have the right papers and the government considered him a migrant worker with *nongye hukou* (agricultural registration). Lately he has been hearing that the rules are changing and that he might be able to stay in Wuzhou without fear of deportation back to his village, but, for now, it remains difficult to get health care from the state due to his registration status. He cannot wait to get those new papers and maybe buy a small place for himself . . . maybe then the company will give him a raise.

Liang's first mobile was a second-hand Bird, which is the Chinese brand. He remembers when he got it, because it was just before the SARS crisis, back in 2003. At first, when the crisis was new, he received text messages about the disease from the government health authorities, and from friends. But soon, the messages from the authorities just stopped; but he kept an active SMS dialogue with friends about what to do, what cities were most impacted, and how to treat the disease if they got it. Some of the texts seemed crazy (Liang got one which said that he could catch SARS just from an SMS message) but he was happy to be able to share any information – it made him feel less isolated during the worst of the crisis, when he would rarely venture outside except to go to work.

These days, he tries to avoid the "crisis" messages. Every now and then, he will receive a message, usually in the form of a joke or some other vague language, protesting against this or that thing that the government is doing. Usually, these are anonymous, but sometimes his friend Shen forwards one of these messages to him. Honestly, Liang would rather keep his head down, at least until that *nongye hukou* thing gets sorted out. You never know who is reading your text messages, and he does not need the trouble. The jokes about some of the officials can be funny, though.

The Bird's battery wore out, and he was making more money, anyway, so Liang wanted to show off a little bit. So, recently, he went to an informal phone shop in the main market in Wuzhou, and bought a second-hand, reconditioned Nokia 8850 for about 250 yuan ($36). It was great value and it is a fancy phone. He has thought about changing his number. It has one eight and a couple of sixes, but no fancy or lucky combinations like his colleague Lei, who has a very lucky number: 1530 871 3488 with its three eights, two of them being consecutive. But Liang does not have an unlucky number either – just a hum-drum number for now. Perhaps after he gets that raise he

will trade up to a luckier number. Or he might get a new handset instead. Shen (the guy who sends Liang the political SMS
messages) has a handset with video, and has a funny video clip
of a dog smoking a cigarette (not to mention some pretty racy
clips of female pop singers); Liang wouldn't mind being able
to get a copy of all of them.

Speaking of girls, Liang is hoping that the handset will
help him find a wife. He has been dialing-in to a "party line"
chat room, and trying to meet a nice girl somewhere nearby.
Unfortunately, it seems like there's only a bunch of other guys
like him on the line, and it's costing him a fortune.

2 Reflections on the vignettes

During the twentieth century, landlines became regular fixtures in domestic, civil and commercial spheres throughout
the developed world. And yet it is remarkable how *unremark-
able* this spread seems to have become (de Sola Pool, 1977;
Fischer, 1992). Many readers of this book will have been born
into homes with landlines in place, and perhaps had a line
of their own even as teenagers. Much of the developed world
has grown comfortable with connectivity at a distance afforded
by the landline network; perhaps many take the privilege of a
voice call almost for granted.

In the developing world, however, the mobile has led to a
second surge of connectivity – every bit the equal of the first
surge brought about by landlines. Clearly, none of the fictional people we described in this chapter take their mobiles
for granted. None had ever owned a landline phone; in each
case, the mobile was their first, primary and basically exclusive means of mediated communication. Only one, Rohit,
uses a PC. Some have purchased second-hand mobiles, while
Prashant still shares a phone with others in his community.
We cannot overemphasize the importance of this basic connectivity to hundreds of millions of people in similar situations.
Like the landline (Boettinger, 1977; Maitland, 1984) and the

telegraph before it, the mobile offers its users the remarkable ability to move their ideas faster than they can walk. Thus it is understandable that many poor households have voted with their pocketbooks, electing to spend upwards of 10 percent of their monthly household income on telecommunications, compared to a level more like 3 percent in more prosperous countries (Gray, 2005; Zainudeen, Samarajiva & Abeysuriya, 2006). The remainder of the chapter will discuss the significance of these vignettes, focusing on the ways in which the mobile is used as a substitute for a landline.

2.1 Mobiles are telephones first

One way to frame the importance of the mobile as a primary means of mediated communication is to focus on the telephone *number* rather than the handset. Granted, marketers (and researchers) are interested in fixing our attention on feature-rich handsets and software, but each new mobile user gains access to the global network of switches, cables, satellites and fiber that connects each telephone to every other. As Colin Cherry said of the landline phone during its centennial year, "perhaps we ought not to celebrate the telephone at all in 1976, but should . . . celebrate instead the telephone *exchange*. It was the exchange principle that led to the growth of endless new social organizations" (1977, 114).

If we think of the number of subscribers, the first decade of the twenty-first century will see the addition of 3 billion new ways to make a telephone call. Thus, our understanding of the impact of mobile communication should be grounded in our understanding of mediated communication, more broadly. Existing research on the role of the telephone in society can help us see continuities and differences between mobile communication and "traditional" mediated communication. This chapter tends to highlight the continuities; the next will highlight some differences.

To make sense of the arrival of mobile telephones in the lives of people like those represented here – Liang, Amihan,

Prashant and so on – we can look at historical and sociological literature on the telephone, some of which has taken the long view of its impacts on the social and economic spheres. It has been difficult at times to step far enough back from the telephone to understand its role in shaping societies. Studies of the landline telephone are relatively scarce, compared to those of the press, the television and, recently, the internet. Nevertheless, scholars developed an intuition about the impact of mediated communication on social organizations, on the workplace, on the home and on human settlements; in short, industrialized societies have been taking advantage of (and been shaped by) the ability to instantaneously communicate at a distance for over a century and a half. As noted in chapter 2, the telegraph (Standage, 1998) and Marconi's wireless radio offered early electric improvements to physically transporting messages, and offered greater distances and better reliability than, say, semaphore signals, smoke signals or lights and mirrors. Quickly, these innovations created crises of control and coordination, as the instantaneous availability of information created opportunities for distributed action, required alignment of time zones and restructured international relations (Beniger, 1986; Carey, 1988; Steele & Stein, 2002).

Though the telegraph was the first modern mediated communications device, the telephone, supported by the exchange system, was the first to make its way into the hands of individuals and small businesses. The year the telephone was invented, 1876, the United States had 214,000 miles of telegraph wire and 8,500 telegraph offices (Aronson, 1977); there were 13 million telephones in operation only 25 years later (Grace, Kenny & Qiang, 2004). By its centennial year, with 228 million lines around the world, its impact on human life was becoming clear.[2] For example, Martin Mayer described how the telephone's intimacy, ease of use and "privity" both allowed the telephone to serve as "an extension of self" (1977, 238) and yet "permits others to impose their time scale upon you" (p. 244). Thinking about the possibility of a wireless phone, he was concerned: "I must say, though, that before I would carry that sort of instrument, I'd insist on a way to turn it off. The telephone may not have reached saturation yet, but some of its users certainly have" (Mayer, 1977, 245).

Indeed, the sheer variety of uses and contexts to which the telephone can be applied undermines the universality of statements about its impacts. To de Sola Pool, "The phone, in short, adds to human freedom, but those who gain freedom can use it however they choose. Rather than constraining action in any one direction, the telephone is an agent of effective action in many directions" (1977, 4). Claude Fischer, focusing on American households prior to 1940, found amplification as often as directional change:

> Americans apparently used home telephones to widen and deepen existing social patterns rather than to alter them. Despite the romance of speaking over long distances, Americans did not forge new links with strange and far away people. Despite experiments with novel applications of telephony, they did not attend concerts, get healed, hold town meetings or change lifestyles via the telephone. As well as using it to make practical life easier, Americans use the telephone to chat more often with neighbors, friends and relatives; to save a walk when a call might do; to stay in touch more easily with people who lived an inconvenient distance away. The telephone resulted in our reinforcement, a deepening, a widening of existing lifestyles more than in any new departure. (Fischer, 1992, 263)

Telephones help people be people and families be families. That does not mean that telephone use is uncomplicated; for example, Rakow describes how the mediated communication afforded by telephones lets women of small children work outside the home. This "remote mothering," at the same time, places new constraints and demands on their time (Rakow, 1992). All of this suggests that any reductionist or purely instrumental lens might not be sufficient to capture what the phone means to people, to discern how and why it is used, or what its "impacts" might be.

2.2 Like landlines, mobiles help support and amplify existing social relationships

Looking back over the vignettes, it is easy to see echoes of this same social amplification effect; *when people have no other telephone, save the mobile, much of that device's use will mirror that of a landline.* Amihan engages in "remote mothering" of her daughter back in the Philippines. Rohit's mom does a bit of remote mothering herself, having purchased a mobile and a subscription for her son in Bangalore. Prashant does not even own his own handset yet, but, when he borrows his friend's mobile, it is to place calls to family members living far from his village. Annette, too, chats with friends and family when she is not working in her restaurant. A study of small-business owners in Rwanda, like Annette, revealed that roughly two-thirds of their total calls were not with business associates, but were rather with friends and family (Donner, 2004b).

Liang's reliance on his mobile during the SARS crisis is also illustrative of earlier uses of the landline; from social support to information seeking, people take advantage of mediated communications to pursue a variety of goals and needs. This is not to say that every telephone call placed on a mobile handset reinforces an existing, strong relationship. Liang, for example, uses the mobile to access chat rooms and to look for romance, and Amihan's daughter Dalisay exchanges text messages with boys much more frequently than Amihan would like. As Fischer's and de Sola Pool's observations about the landline make clear, people can use mobiles to pursue relations of various kinds, durations and properties. Chat lines and romantic texts are common to mobiles (Pertierra, 2006), but they have their roots in landline features, whether paid services or even party lines. Despite the occasional "moral panics" in the mass media about the role of mobiles in romantic relationships, it may be overly reductionist to suggest that mobiles *cause* teen romance in the Philippines or China or India. Rather, as we'll cover in more detail in the next chapter, mobiles present new challenges around age-old issues; like mass adoption of the automobile in 1950s USA or the placement of landlines in the bedrooms of American teens a few decades later, the availability of mobiles has been a boon to young people looking for companionship, and a source of concern for their parents (Donner, Rangaswamy, Steenson & Wei, 2008).

Thus, the content, purpose or impact of the teenagers' calls might not be very different from what we might see with landlines. However, thanks to the different economics of mobile coverage, to advances in electronics, miniaturization, computation and connectivity, and to changes in the global economy which have given each of them greater purchasing power than their parents or grandparents had at their age, they are able to do via a mobile handset what other people have been doing for over a century – they can make and receive calls which can help them pursue the lives they wish to pursue.

2.3 Like landlines, mobiles help people be productive and pursue their livelihoods

There is a second discussion, however, in which the instrumental, directed benefits associated with telephony have been of central concern. We can also look at the specialized research on the role of telephones in economic development. For at least half a century, policymakers and researchers have been amassing specific evidence about the role which telephones play in supporting growth and alleviating poverty; their insights about markets, security, isolation and coordination ring truer than ever, as mobiles have increased the availability (and value) of the telephone networks around the world. Telephones allow individuals and organizations to gather and exchange information about prices, transportation, medical services, emergencies, new market opportunities and so on. These calls are critical inputs not just to high-tech "information age" businesses but to agriculture, manufacturing, services and trade (Hardy, 1980; Hudson, 1984, 2006). Households clearly value telecommunications services, being willing to forgo expenditures on other things to pay for telephony (Torero, 2000).

Governments, too, rely on telephony for everything from public health surveillance to taxation.

Debates remain about measuring the exact magnitude to which, and causal processes by which, the telephone or telephone penetration contributes to growth and economic development (Roller & Waverman, 2001). But, as we have noted in chapter 1, the Maitland Commission (Maitland, 1984) pointed to a major discrepancy between access to telecommunications services in the developed and the developing worlds. Economists working with the World Bank have gathered and integrated hundreds of studies about telecommunications and economic development. They summarize telephony's contributions as including: better market information; improved transport efficiency and more distributed economic development; reduction of isolation and increase in security for villages, organizations and people; and increased connectivity to (and coordination with) international economic activity (Saunders, Warford & Wellenieus, 1994, 23–9).

These same four major categories of impacts of telephony apply to mobiles as well, particularly and most visibly in the developing world, but also elsewhere, where billions of mobiles can be used in the same instrumental, productive ways as landlines. The boom in mobile telephony has brought a major new wave of expansion in the telephone network. What we see now is a second wave in this same process: previously excluded neighborhoods, rural areas, whole regions and even whole countries (Somalia) "suddenly" have telecoms. All told, at a macro level, one econometric study suggests that, among low-income nations, an increase in mobile penetration of 10 lines per 100 is associated with a 0.59 percent increase in GDP per capita growth (Waverman, Meschi & Fuss, 2005, 18). Much work remains to be done before we have an exhaustive, comprehensive or universal model of the causalities underlying this observed correlation, but the variety of mechanisms is illustrated in the vignettes, which echo many of the themes from the Saunders, Warford and Wellenieus list. Using this list, we see the following.

2.3.1 MOBILES SUPPLY PRICE AND MARKET INFORMATION

Mobile telephony allows people to place orders, arrange shipments, check prices and make deals. In what is perhaps among the clearer and more elegant studies of the economic impacts of mobile communication in the developing world, Robert Jensen (2007) demonstrates the impact of the introduction of mobiles (and their ability to convey price and demand information) on the price of fish. By tracking the price of fish at weekly intervals over five years at various port markets in the Indian state of Kerala, Jensen is able to demonstrate how the price of sardines stabilized almost immediately after the arrival of mobiles, reducing waste, and allowing fishermen to earn more money on average from their catches. In one sense, the critical missing element was basic information about prices and demand – the kind of information which can generally be conveyed by landlines (e.g. Eggleston, Jensen & Zeckhauser, 2002). But of course, in another sense, no fisherman was ever going to benefit from landline telephony – the nature of their days cruising the near-shore seas means that only a mobile/ wireless solution could provide the connectivity they needed.

In the vignettes, we see Prashant, who is taking advantage of connectivity at a distance when he checks prices or his balance due to the sugarcane collective, using a text messaging system instead of voice calls to gather the information he needs.

2.3.2 MOBILES HELP IMPROVE TRANSPORT EFFICIENCY AND DISTRIBUTE ECONOMIC DEVELOPMENT

If Annette had a landline at the restaurant, customers could call her there to place orders and check on the specials of the day; but she does not. Her mobile is the tool that allows her to "micro-coordinate" with customers. Annette's customers at the restaurant in Kigali are also using mobiles to ask her about her products: to place orders for lunch and to see if food is still available. Some of these interactions are so routine that they can be done with a "beep" – an intentional missed call that conveys pre-negotiated information without the price of a voice call or text message. When her customers place a call, they can save a trip; and even when they decide to arrive, the food is hot and waiting at the table for them. When Prashant orders fertilizer or a new tool from his cooperative, or from another vendor, using the mobile, he is saving time, swapping calls for trips; it also creates a rural–urban linkage, and the tool vendors can serve a wider geographic sweep of customers more effectively thanks to the benefits of coordination at a distance (Overå, 2008). Other farmers in India can activate irrigation pumps in their fields via SMS, again saving a trip to the fields. These little things, repeated billions of times by hundreds of millions of users, amplify, strengthen and diversify the opportunity for economic and civic interactions between people and organizations.

Annette's small business is benefiting from her access to the mobile; so too is Prashant's. However, these examples of how mobile phones impact the livelihoods of many people in the developing world do not imply that *everyone* benefits from mobile communications in this way (Miller, 2006). Not every business will improve thanks to access to mobile communications. Nor does everyone use a mobile to start or improve an enterprise. Indeed, not everyone has an enterprise to improve. Nevertheless, the utility of the mobile phone to some businesses and livelihoods is clear. Hundreds of millions of households throughout the developing world rely on small and informal enterprises to earn their livelihoods, hundreds of millions more rely on small-scale agriculture – thus, the capacity of the mobile phone to help match buyers and sellers, to reduce price uncertainty and to lower the costs of transactions is helpful indeed.

2.3.3 MOBILES REDUCE ISOLATION AND INCREASE SECURITY
Beyond strengthening of domestic markets, mobiles, like landlines and telegraphs, help tie together rural and urban areas, allowing greater coordination of government services.

Even when living in an urban area, Liang found that his government communicated to him via mobile, sending him SMS messages with updates and instructions around the time of the SARS crisis.

Public health projects are illustrative here (Donner, 2004a). In Africa, companies like Voxiva (Casas & LaJoie, 2003) and Cell-Life have developed public health informatics platforms which allow health ministries to send queries, gather statistics and coordinate activities with rural health clinics, many of which lack electricity or a landline connection. From gathering data on the status of stocks of antiretroviral drugs to treat HIV/AIDS to empowering nurses to offer better prenatal and antenatal care, the connectivity offered by mobiles has helped cut into the isolation that has separated villages from their governments. Sometimes, advanced software solutions are not even required; Idowu, Ogunbodede & Idowu (2003) describe how doctors in a Nigerian hospital use their own personal mobiles as their paging and voice mail system, because the hospital lacks a formal (and expensive) voice mail system of its own. Rural residents can use mobiles to call for ambulances in emergencies (Souter et al., 2005).

2.3.4 MOBILES FACILITATE INTERNATIONAL ECONOMIC LINKAGES
The higher cost of international calls might prevent many small businesses from using mobiles to seek international markets (Molony, 2006); conversely, larger businesses already have landlines. So the link between mobiles and international trade may not be as strong as that between the landline or telegraphs and trade.

However, Amihan's and Annette's stories illustrate how the accessible, affordable connectivity provided by mobile communication helps some families stay together over long distances, even when the reason for that distance is the pursuit of economic opportunity (Paragas, 2005). By some estimates, there were over 200 million overseas diaspora workers in 2007, sending roughly $320 billion annually

(GSM Association, 2007) to their families back home. Others are migrants (like Annette), permanent settlers in one country from another. Countless millions more, like Liang, have relocated within countries, moving from rural areas to urban ones in search of better jobs or a livelihood. Undoubtedly, mobiles help them stay closer to their families over distance – might this connectivity help them stay away longer, does it help them spend less to stay connected, or do they perhaps just feel more connected when working away? In any case, it is difficult to draw a bright line between economic and non-economic use of the mobile. It is not uncommon for family members living overseas to buy mobiles for their families back home (Tall, 2004); the conversations carried by those mobiles will mix social and economic subjects.

2.4 Mobiles and the digital divide

The spread of mobiles in the developing world has not gone unnoticed by policymakers, nor by the press. Cover stories in daily newspapers regularly trumpet how mobiles are closing the "digital divide." In a manner of speaking, they are correct; people who previously had little access to affordable telephone services now have that access, and millions are taking advantage of it. But the term "digital divide" is slippery. It originally described a socioeconomic gap between computer users and nonusers in the United States. A more direct antecedent might be the "universal service" obligations in the US and European landline rollouts; universal service obligations are being re-worked in the mobile sphere as nations allocate the valuable spectrum – in India, for example. There is no doubt that governments, policymakers, etc., have a role to play, and that ICTs can be more or less "pro-poor" depending on how they are configured in society (Forestier, Grace & Kenny, 2002).

But mobiles are no more of a cure-all than any other technology. In livelihood studies, for example, some point out how access to / use of mobiles may not overcome other structural factors. Not everyone may profit equally from mobiles – traders, for example, may do better than producers (Jagun, Heeks & Whalley, 2007), and voice calls may not undo or replace time-tested trustworthy face-to-face relationships (Molony, 2006). Mobiles may help close an access gap, but do not (a) immediately re-structure human affairs in a more "just" way, nor (b) provide the motivation, affordability or even technical capacity to access and use advanced informational services like GIS, medical imaging, spreadsheets, etc. (Donner, 2008b).

Voice calls and text messages remain the bulk of most people's mobile behavior. Indeed, for every mobile internet session or use of advanced services there are thousands, if not millions, of SMS messages and voice calls. This said, some people are experimenting with more advanced and varied functionality, from cameras and video to music, internet and e-mail. Still others use simple text or voice interfaces to access "advanced" services sitting on the servers, as in the case of m-banking or m-government. One example is the mobile-enabled transactions and membership database managed by Prashant's sugarcane collective. Agricultural exchanges and m-transactions hubs like Senegal's Manobi and Bangladesh's Cell Bazaar are springing up around the developing world, letting farmers conduct business at a distance, in ways previously restricted to those with PC or dial-up internet access (infoDEV, 2004).

Another example is Amihan's remittances, which she sends back to Roxas using an m-banking system. There are far more people in the world with mobile phones than with bank accounts (Porteous, 2006), but, increasingly, the mobile can be used as a way to store currency and to pay for transactions, even for people without formal checking accounts or credit cards (Donner & Tellez, 2008). In Kenya, one such m-payments system is M-PESA, which has signed up over 2 million users in the first 18 months of operation. Users no longer have to wait in line at banks to send money from one part of the country to another; they simply "cash in" at an agent's shop and then send funds to friends or family via SMS (Morawczynski & Miscione, 2008). We are just scratching the surface of what m-commerce and m-services can provide; but in the last few years, innovations have begun to come from the developing world as well as the developed. In the case of m-banking, the Philippines, South Africa and Kenya are leading the way (infoDEV, 2006).

Returning to the "digital divide" as a matter of internet access, rather than telephony, there's room for optimism. Recent improvements to web browsers on high-end phones will begin to trickle down to middle-range phones, while the proportion of even the most basic phones which support some simple (if not graphically beautiful or particularly speedy) data transfer will continue to rise. Data access will most likely improve and, perhaps, become more affordable in many markets. That said, neither the research community nor industry knows for sure what proportion of the world's population currently not online will elect to use the mobile as their primary (or only) internet access device. The problem is not with the march of technological progress and adoption, but rather with the term "digital divide". Indeed, while the "entry-level" phone of 2015 will probably contain many features of today's top-end models, the top-end phones of 2015 will also have shifted, perhaps to faster networks, better displays, better or completely redesigned interfaces, more processing power, etc. As long as there are differences in income, skills, literacy and simple personal preference, there is likely to be some kind of "digital divide."

3 Summary – mobiles are another step towards a global information society

We will see the addition of over 3 billion new mobile telephone lines during the first decade of the twenty-first century. Many of these lines are in the developing world, where hundreds of millions of people have access to affordable telephony for the first time. Through a series of fictional vignettes, we have outlined how some of these new users – people like Prashant, Amihan, Liang, Rohit and Annette – use this new communications possibility on a daily basis. We used their stories to illustrate two great continuities between mobile phones and earlier mediated communications, particularly the telephone. Like landlines, mobiles support and amplify social relationships. Moreover, like landlines, mobiles help people and organizations be more productive, by allowing the quick transmission of information, by substituting for journeys, by increasing security and by allowing for coordination of events at a distance.

This is not to say that mobile phones are not something new. The next three chapters will deal directly with the new norms and challenges presented by the arrival of widespread mobile communication. But we do want to close this chapter with a reminder that mobiles are a force of continuity as well as change. Unlike CB radios, for example, mobiles connected to the landline telephone exchanges almost from the start, which made both mobiles and existing landlines more valuable.

Broader, existing descriptions of the role of mediated communications in modern society are an important starting point for understanding many of the impacts of mobile communication to date. We have highlighted a few such descriptions in this chapter, particularly Fischer on the history of America's early adoption of the landline and the birth of the "social call" in pre-Second World War America, and Beniger on the century-long control revolution in the productive sphere and the arrival of real-time coordination of activities at a distance. As mobiles are sweeping the world, bringing huge spikes in connectivity and telephony-per-capita to the 4–5 billion people who started the twenty-first century without a telephone, the patterns identified by Beniger and Fischer, and by many other students of the landline, remain valid.

Moving to an even higher level of generalization, we can recall the general formulation of Castells' "network society." To Castells, the networks of informational exchange are so central to the organization of our productive and social spheres that they have come to define our age. These networks began with the telegraph, took root with the landline, and boomed in complexity and power thanks to everything from satellites to television to the internet itself. Of course, the economy of the early twenty-first century is not all virtual. People grow food, build houses and dig mines as they have done for centuries. Nor have our social interactions gone all-mediated. The vast majority of people (even Facebook and World of Warcraft addicts) have families, friends, coworkers and neighbors they see at least sometimes, face to face. But "Networks," argued Castells in 1996, just before the mobile explosion, "constitute the new social morphology of our societies, and the diffusion of networking logic substantially modifies the operation and outcomes in processes of production, experience, power, and culture" (Castells, 1996, 469). Connections to these network flows, of media and telecommunications, of highways and international trade, order and re-shape all the stuff of non-electronic everyday life. In that sense, mobiles are *extensions* of these networks, not replacements or disruptors of them, a point Castells and his colleagues emphasize in their review text, *Mobile Communication and Society* (Castells et al., 2007, 258).

At the same time, mobiles are remarkably numerous, offer greater mobility, increased and ubiquitous processing power and flexibility, and are often linked to people, rather than physical places. For these reasons, the devices do present different challenges to users, non-users, organizations and societies – ones that even the best students of telecommunication, like Beniger and Fischer (or the members of the Maitland Commission), were largely unable to anticipate. We will examine some of these challenges in the next chapter.

CHAPTER FOUR

Mobile communication in everyday life: new choices, new challenges

1 Introduction

In chapter 3, we examined the role of mobile communication for those who had never, or perhaps only rarely, used a landline telephone, and argued that, in many ways, the mobile phone needs to be understood as a telephone first – we emphasized that the main benefit of mobile telephony to many people is the "telephony" part, not necessarily the "mobile" part. That said, even amongst the vignettes of the previous chapter, we could see examples of how the mobile phone presents new choices and new challenges to its users, compared to the traditional landline. In this chapter we will draw a series of fictional vignettes describing mobile phone use for those who use mobiles as complements to landlines. Two major differences stand out: the convergence of the voice features with other media, features and functions, from cameras to music players to the World Wide Web, and, second, the wirelessness and portability of the device, which has made so many of its users reachable anywhere, anytime.

This second set of vignettes will further underscore the breadth of use patterns around the world. Differences in the way telephone companies commercialize their offerings, differences in the technologies used, differences in the way governments have regulated the industry and differences in everyday users have resulted in quite different behaviors. For example, *keitai* in Japan are both powerful and "cute," and are the device of choice for most Japanese. In the USA, "cell phones" are bundled with postpaid family plans and corporate accounts, and compete with instant messaging, e-mails, laptops and landlines for the attention of their users. In sub-Saharan Africa, prepaid top-up accounts have given millions of people their first real access to telephony. They need only to have a small SIM card that they can insert in another person's handset. Within societies, the differences are just as great; life stages, socioeconomic status, vocations and simple personal preferences create considerable variety in the ways people use mobiles. The following stories only begin to capture the variety of ways people choose to use their mobiles in everyday life.

1.1 Mette, Oslo, Norway

Mette, a 47-year-old woman who lives in Oslo. She works in hospital administration. (Based on interaction with Marit Sandvik.)

> **Phone type:** Nokia 3100
> **Subscription type:** prepaid "top-up" subscription
> **Average monthly expenditure:** $20 (about 1 hour's work)
> **Hours of work to buy her mobile phone handset:** $120 (about 6 hours' work)
> **Most common use:** voice calls to her family, with some texting

Mette is a 47-year-old woman who lives in Oslo. She is an administrator at one of the large hospitals in the city and she lives with her husband Roald and their two children (Inger aged 15 and Per aged 12). They live in a comfortable section of the city that is not far from the forest, with a view of the Oslo Fjord. As with many families in Oslo, they have only one car and indeed rely on public transport for most commuting.

Each member of the household has a mobile phone. Mette has an older green and white Nokia with a black and white screen and the basic functions. She has a perverse pride in having an old telephone when many of those around her have new sleek devices that have tons of functions – which they never use. By contrast, her mobile phone is robust and easy to take care of. Her children make fun of it as being a museum piece. Their daughter Inger has a more advanced mobile telephone that allows her to play music and take photos, and indeed her mobile phone and her MP3 player are usually with her. She is proficient at sending text messages and maintains a seemingly constant stream of interactions with her school friends. Per has a mobile phone, but often forgets it and it is not often charged up. Roald has the most advanced mobile phone that is provided by his employer. He has a device that allows him to check his e-mail and surf the net in addition to a broad range of other functions – he uses e-mail the most. The family also has a traditional landline phone that is in frequent use.

Mette was late in adopting the use of the mobile phone. Indeed, it took a small conspiracy five years earlier between Roald, Inger and Per to put one in Mette's hands. She had resisted getting a phone since she felt that it was not necessary, just a fad and only a waste of money. During a Saturday shopping expedition, the three had decided on the model and had bought it. In addition, they bought her a prepaid "top-up" subscription. With this type of subscription, there was no monthly charge, but the price of each call and text message was somewhat higher than for subscriptions with a fixed monthly fee. This allowed her to prepay for her use and, when she had used up her credit, she could "replenish" her account. Mette was not a part of the decision-making process. At that time, the mobile phone cost about 1,000 Norwegian crowns (that is, a bit more than a half a day's wages). The prepaid access can be bought in various units ranging from 50 to 500 crowns.

At first Mette rarely used the phone. She did not see herself as the type of person who needed a mobile phone. Only a few people had her telephone number and she almost never had her phone turned on and with her. It was usually either Inger or Per who might call her when they were done with one of their free-time activities.

At one point, her mother – who lives in a small village on the west coast of Norway – was sick and admitted to the local hospital. During that period, Mette carried the phone with her to be in touch with the hospital and to be available to her father and her brother. After this crisis passed, the mobile phone reverted to its more restricted role in Mette's life. However, the situation pointed to the benefit of being available as needed, and the lesson was not lost on Mette.

A couple of years after Mette got her phone, Inger entered middle school. In addition to being located somewhat further away from home, the transition also marked Inger's growing social engagement in her peer group. Inger became involved in the local cross-country skiing club. She competes in different meets held in the greater Oslo area. In addition, she is often invited to the home of one of her friends. Per often plays soccer after school and Roald was sometimes away on business trips. Given all these diverse trajectories, the coordination of the family became increasingly complex. Mette soon saw the convenience of having a mobile phone since it facilitated the coordination of different family errands. Roald could call from the store to ask whether they had enough cat food or whether he should also get some clothes at the cleaners. Inger could call and ask permission to visit a friend after skiing practice, and Mette was available to Per when he was done with school.

Both Inger and Per became proficient at the use of texting. To be sure, Inger was entering into that phase of life where she sent and received what Mette seemed to think was dozens of messages every minute. Mette and Roald soon started receiving text messages from Inger. It was, in effect, Inger who taught them how to tap out the messages on the keypad of the mobile.

As time went on, Mette became more proficient in her use of texting and began to receive messages from many sources: from her colleagues at work when different crises erupted; from the children when she was at work when they had things that needed to be dealt with; from Roald; and on occasion from, for example, a repair shop when the item being repaired was ready to be picked up. Sometimes it seems to be a little too much.

While she lags badly behind her teenage daughter in the use

of the device, it has become a standard part of her kit. More recently, the family has changed to a "family subscription" that allows free calls amongst their phones. This, along with Mette's growing reliance on the device, has meant that the mobile phone has become increasingly central in her coordination of the family.

Mette generally makes one or two short calls and perhaps sends a text message each day. These generally deal with arranging to pick up one of the children or deciding on a meeting place while on the fly. The mobile and texting are also used to keep a cautious connection with Inger who is starting to explore the world of boys, parties and the pursuits of adult life. Mette and Roald do not call Inger often when she is out in the evening. They do, however, send her a discreet text message when it is starting to get late. Thus, rather than humiliate her in front of her friends, they engage in a quiet "background" type of interaction that allows them to keep the information channel open.

Mette does not use any advanced services on her phone and indeed the device is so old that it is not possible, for example, to surf the net or to use WAP services. Mette still uses the landline phone if she wants to have a longer phone conversation with either one of her friends or with her mother.

1.2 Mika, Nagoya, Japan

Mika, 16, is a teen who lives with her parents in Nagoya, Japan. (Based on interaction with Satomi Sugiyama.)

Phone type: a beige FOMA SO703i flip phone with a "Miffy" strap and a stuffed bear charm and a photo of her friends attached to the back of the device

Subscription type: a "family plan" where her parents pay for 4,000 minutes and 1,000 *meiru* (text messages) that Mika, her 14-year-old brother Kano and their parents must share. Her family pays about ¥13,000 per month (about $75) for their service.

Average monthly expenditure: her parents pay for her use as part of their family plan

Hours of work to buy her mobile phone handset: with the use of bonus points, it took Mika's father about 4 hours' work to pay for her new handset

Most common use: mostly texting but some voice calls

Mika is a 16-year-old teen living in Nagoya. She attends Toho High School and although they are a couple of years in the future, she spends a large portion of her day studying for her college exams. She is good in literature and is among the best in her class at learning languages. Math, however, does not come easily, so she needs to spend a lot of time working on that. On most days, in the evening after school, she attends *juku* or a "cram school" that focuses on helping her with the entrance exam for a university education. She attends this with some of her friends and so, in addition to the hard work, she gets to spend some social time with her peers. She also takes piano lessons through an after-school program.

Mika has had her phone for the past three years. Her parents bought her a phone when she started in middle school. Getting to the school involved taking the tram and so her parents felt that they would have better contact with her if she had a *keitai* or mobile phone. Now she sends *meiru* to her mother when she is going to come home late from her evening classes. On one occasion, Mika accidentally forgot to bring her homework to school. She was able to send a text message to her mother, asking her to bring it to her at school. Her mother was not very happy about having to do it, but for Mika it saved the day. In addition to using her mobile to keep in touch with her parents, she also is quite active in sending and receiving text messages with her friends.

Not long ago, she was able to get a new *keitai* or mobile phone. She had accumulated enough points to get a FOMA SO703i. The phone is a sleek Aqua-White "flip phone." The points are a type of loyalty scheme. Users receive 1 point per

100 yen of use, and they can also receive bonus points in some cases. These points can be used in the purchase of a new phone, where 100 points can result in a saving of 100 yen. (A yen is worth about US$0.01.) Mika had accumulated 5,000 points at the time of her purchase, and the phone cost 18,000 yen. Her father contributed the remaining 13,000 yen since she had worked so hard at school.

She spent several weeks thinking about which phone to buy. There are so many different phones she needed to think about. Some are boxy in bright colors, while others are round and sleek. In the end, she and her two best friends, Natsumi and Yukako, had a long chat about which phone was best. They all agreed that the phone she chose is a very feminine phone. Natsumi had just gotten a new phone that she was only moderately happy with. She liked the design of it but the keys were difficult to push when composing text messages. Yukako had an older phone and wanted to buy a new one. She was the one who had noticed that particular FOMA phone. She was working on saving enough points to get a black version of the same one. Her friend Hoshi pointed it out in a store window one day and Mika agreed that it was a nice design. She did not like the color, however. When she found out that there was a beige version of the phone, Mika thought that that was exactly what she wanted. A week later, she had bought the phone. As had Natsumi and Yukako, Mika also added a strap showing her favorite Miffy cartoon character, and she also hung a small stuffed bear charm on her *keitai*. The bear is the same one that her friends Nayoko and Hoshi have on their phones.

Mika likes to meet her friends near the central train station when they have free time. They use *meiru* to agree about where to meet. Often they meet at JR Takashimaya and chat, or go to a small crepe restaurant. A week ago she was able to spend a Saturday with her two friends, in addition to two boys, Dai and Takuya. All five of them went to a *purikura* (photo machine) and had a photo taken. The photos were printed on thumbnail-sized self-adhesive stickers. Now all of them have a photo sticker mounted on their phones. Mika is secretly in love with Dai and the photo is a good way for her to keep him close without being obvious about her special feelings towards him. She was so excited about getting the photo that she spent the whole train ride home sending *decomail* (decorated text messages, with various colors, fonts, backgrounds, attached photos, etc.) about it to Natsumi and Yukako.

Like most of her friends, Mika has an I-mode subscription. The most frequent service she uses is the texting function that is also accessible from a PC. There is a special I-mode button on her phone that sends her to the start menu. Once there she can enter the I-mode sites. The *meiru* function is a part of I-mode. In addition, she can check on the weather and sometimes she uses the train schedules that are posted there. The sites of her favorite J-pop singers, Koda Kumi, Namie Amuro and the band Mr. Children, are also part of the I-mode function. Mika often visits these sites via her mobile phone. In addition, she occasionally downloads some of their songs. She has one of Koda Kumi's songs as her ringtone. At one point, Mika saw a TV broadcast of Namie Amuro via a friend's mobile phone. Now she secretly would like to use that service. She hopes that if she gets good grades in school her parents will let her subscribe to it.

Since her subscription is a part of her family plan she does not pay for her use. That is taken care of by her parents. She needs to be careful, however, not to send too many messages since her parents would get upset and she would get into trouble.

1.3 Alberto, Milan, Italy

Alberto, 45, is a businessman in Milan. He owns and operates an agency that helps stage fashion shows. (Based on interaction with Leopoldina Fortunati.)

Phone type: Prada/LG smart phone
Subscription type: postpaid subscription paid through his job

Average monthly expenditure: $250 (paid for through his company as an operating expense)
Hours of work to buy his mobile phone handset: 2
Most common use: voice calls, texting, e-mail, scheduling, photos

Alberto is a 45-year-old businessman who is involved in the fashion industry in Milan. He owns a small agency that produces fashion shows. His agency assists the fashion houses in finding locations and staging their shows during the Milano Moda Donna (Milan Fashion Week). He has five people employed in his agency and, depending on the needs of his clients, he can call on other individuals and groups for their services. He currently has three clients for the upcoming show featuring Fall fashions: Prada, Missoni and the British company Burberry. He lives with his partner, Tito, in an apartment north of the city center. His partner is a doctor at the university hospital located near the center of the city.

The mobile phone has become a central tool in Alberto's job. In his daily work, he often must move about the city to different locations where his clients want to stage fashion shows. In addition, he needs to be in contact with different jobbers and production assistance personnel. He has to interact with clients, stage designers, architects, lighting and sound specialists, electricians and many other groups who are involved in the staging of the fashion shows. He uses his mobile phone almost continually. He is in contact with different individuals via the use of voice and texting, and he has set up his phone so that he can read his e-mail. He keeps his calendar of events on his phone along with his "to do" list. This is automatically synchronized with his PC via the mobile phone so that he does not need to go through the process of physically synchronizing the two. The mobile phone is used for coordinating the different phases of his work. The design of the show is determined in a series of meetings with the client and the architects and designers. Alberto and his colleagues then work out the logistics of realizing the design and contracting the different specialists who will do the actual work.

In this process, the mobile phone has become progressively more important, as neither Alberto nor the different specialists are tied to a physical location. Rather they work at many different locations around the city. The coordination of these people has become increasingly possible since literally all the individuals upon whom Alberto relies also have their own mobile phones. All the workers, the suppliers and the professionals can be individually contacted. This has been a major change in the industry since Alberto started in the mid-1980s. At that time there were only a few of the richest individuals who had a "car phone." Since that time, the mobile phone has become as essential to Alberto and his colleagues as the hammer is to the carpenter. He literally cannot imagine doing his work without it.

In addition to the coordination of logistics, Alberto also employs the camera function to note promising locations and to record the progress of different projects. He takes photos of different possible locations when he is out in the city. He uses these in discussions with clients when thinking about new projects. He also takes photos of the work being done in case there is a question as to who is responsible for delays or problematic issues, and when, for example, the plans of the designers and architects do not mesh with the situation at the locations. In cases like this, Alberto often uses the camera phone to send his colleagues photos asking for their suggestions. Just a week ago he was able to resolve a difficult situation in this way. The architect had drawn an entrance to the catwalk that was too narrow for the needs of the client. Alberto took a photo of the entrance and sent it to the architect, who immediately saw the problem and was able to make the appropriate adjustments.

Alberto also uses the internet function on his phone when he needs to look up information on the materials being used for the construction of the settings. He has used it, for example, to check the prices of lighting fixtures against those suggested by a supplier (the supplier was exceedingly overpriced) and to find out about the availability and the specifications of a special flat screen TV that was used in a recent show (the one the architect had suggested was too small). Alberto can do these things on the fly. He does not have to wait to get back to his office; rather, he can use a few minutes to get this information via his mobile phone.

In addition, Alberto uses the mobile phone to interact with Tito. Since both of them lead hectic, and to some degree unpredictable, lives, they need a secure communications link that allows them to plan who will be at home to make dinner, etc. This usually involves a "coordination" call immediately after lunch and another towards the end of the day. The device allows Alberto and Tito to deal with domestic matters while they are both at work. At the same time, it also allows their working lives to intrude into their private time. Both Alberto and Tito try to limit the degree to which they allow this to happen. In general, they manage to keep work at bay. However, as a deadline approaches or if there is a particular emergency that arises, they are accessible via their mobile phones. Thus, they see the device as being a mixed blessing.

Alberto has a high-end "smart phone" that is paid for by his company. He changes telephones relatively often. If he finds one that he likes and that looks stylish, he might keep it for a year. However, he has had several phones that did not fulfill their promise. These were often replaced with another within a few weeks. He is currently using a Prada phone, somewhat out of loyalty to his customer. He has a traditional postpaid subscription that allows him use of both the telephonic and the data functions of his phone. Tito also has a postpaid subscription.

1.4 Thomas and Flores, Valparaiso, Chile
Thomas (69) and Flores (66) are retirees living in Valparaiso, Chile

Phone type: Nokia 3310
Subscription type: prepaid "top-up" subscription
Average monthly expenditure: $50
Hours of work to their mobile phone handset: 10 (using their pre-retirement income levels)
Most common use: a safety link and for contacting family

Thomas and Flores are retirees living in Valparaiso, Chile. He is 69 and she is 66. Thomas and his wife Flores worked until they were 65 and 62 respectively, but for the past four years they have been retirees. Thomas was a purchasing agent at the University of Valparaiso. He had responsibility for administering large contracts with organizations who delivered food and maintenance services to the university. Flores worked as an accountant for a shipping company that had its home office in Valparaiso.

That was the past. Now Thomas and Flores lead a relaxed and somewhat staid life. They have two grown children. Their son, Pablo, lives several hundred kilometers south of Valparaiso in Valdivia with his wife and their three children. He works for one of the larger wine producers there. Thomas and Flores' daughter, Maria, lives in Valparaiso with her husband and their two children.

Thomas and Flores purchased a single mobile five years ago, in anticipation of their retirement. They reasoned that, in the coming years, they would not have their job-based landline numbers. A mobile phone would help with their availability. In addition, they planned an extended visit to the Torres del Paine National Park and the area around the Magellan straits in the southern part of Chile. A mobile telephone would allow them to get assistance if they had problems with their car, and to contact their children from time to time. If the truth is told, Thomas was the most interested in getting the phone. Flores was somewhat more reluctant. They have a prepaid "top-up" subscription. The minutes cost more than if they subscribed to a calling plan, but they use the phone so rarely that it does not make sense for them to pay a fixed fee every month regardless of their use.

Thomas often has the phone with him when he is away from their home. Flores can then call him just to check in, or to ask him to pick up small items at the store on his way home. They do not use texting. Their daughter Maria sent them a text message several years ago. When it came, Flores was at home alone and the phone started to ring, but when she tried to answer it as though it was a voice call, nobody was on the other end of the line. She tried to explain it to Thomas when he came home, but neither of them could make sense of the situation. It was only after their daughter called and explained that they understood that the telephone had a texting function. They have not bothered to use it since then.

Thomas has been active in the local soccer team. As a young man he aspired to play on the team the Santiago Wanderers, known locally as the "Vagabundos." He never realized this dream, but he did play in the secondary teams and he was also a coach on his son's team when he was growing up. After Pablo went to the university in Santiago to study agriculture, Thomas became a referee. He has continued working with the team by coaching the youngest players aged seven to ten, particularly since two of his grandchildren have started to play. One of his great joys in life is seeing them play with the local team. The soccer pitch is only about a ten-minute walk from his home so he is often there to watch the younger people practice and play. In particular, he thinks that his youngest granddaughter shows great promise. He is frequently at the practice field when she is either training or playing a match. Sometimes Flores calls Thomas on their mobile phone there when it starts to get dark and she starts to wonder when he will be coming home. This reassures her that he is OK and allows her to gauge when she should start preparing dinner.

One day Thomas was on his way home from the soccer pitch after dark. He was carrying some equipment and he stumbled on the curb. He fell and twisted his ankle in addition to getting a couple of scrapes. He remembered that he had his phone and so he was able to call Flores, who in turn called their daughter Maria. She drove down to where her father was and took him to the emergency room. Luckily, there were no serious injuries. Now, however, Flores insists that Thomas has the phone with him whenever he leaves the house.

In addition to their everyday life, Thomas and Flores visit Pablo and his family a couple of times a year. It is a long drive to Valdivia and they like to be able to stay in contact via the mobile phone in case something should happen.

1.5 Stan, Los Angeles, California, USA
Stan, 20, is from Los Angeles. He is a university student with a part-time job. (Based on interaction with Casey Jenkins.)

Phone type: Samsung "slide" telephone
Subscription type: family plan with one of the major companies
Average monthly expenditure: $250 for the four users in his family (of this $40 comes out of Stan's pocket and the rest is paid via his parents' family plan)
Hours of work to buy his mobile phone handset: 13 hours to replace his current handset if he were to commit himself to a new two-year contract. To buy a new phone with some neat functions outside his contract would take about 60 hours of work.
Most common use: voice calls, texting with an occasional photo

Stan is a 20-year-old male from Glendora, California, a suburb of Los Angeles. He is in the process of moving out of his parents' home as he is a sophomore at the University of California at Santa Barbara. He lived at home last summer, but for the coming summer he has been offered a job as a river rafter for an outfitter in Arizona. Thus, he still has his parents' home as a kind of base, but he is spending less and less time there.

Stan always has his phone with him and he is a persistent texter. He has a circle of friends with whom he is in contact and he is part of a mountain biking club on campus that regularly makes trips to different locations around LA and southern California. He receives messages from his friends with regards to when and where to meet, who will be driving, etc. He has a regular digital camera, but he often uses the camera function on his phone to take snapshots during these trips. He almost never sends these as messages, but later downloads them to his PC for posting on the mountain bike group's web site.

He got his phone when he was still living at home. Several of his friends had begun to get phones and his parents each had one for coordinating their trips to the store, picking up the children, etc. Stan had begun to drive and had an old banger of a car. Late one evening, on his way home from a part-time job, his car stalled about 5 miles from home. He did not have any change to make a phone call and, besides, there was no phone booth in the area. It was so late that most of the stores were closed and so he was marooned. He tried to hitchhike, but not many cars were driving through the area and those who were there were not willing to pick him up. A bit later on, it started to rain. By the time he got home it was late, he was drenched to the bone and his parents were, by turns, anxious that something had happened and angry with him for not getting in touch. A month later, his father's mobile phone contract came up for renewal and they decided to get a family plan. Stan and his younger brother each got phones and, for about $250 per month, they got a plan with 2,000 minutes and no text messages. Incoming and outgoing calls both counted against this "bucket" of minutes.

At first, Stan did not send any text messages. This is because there were free calls within the family and he used PC-based instant messaging with his friends. In addition, it seemed nerdish to "multi-punch" in messages. At about this time he hooked-up with his girlfriend, Vickie. They discovered that the nice thing about texts was that they are not intrusive. If they turned off the sound on their phones, they could text one another in class or late at night. After a month, however, Stan's dad showed him the bill for the text messages. Since they did not have a contract that included texts, they had cost 15 cents apiece. The 240 that Stan and Vickie had sent one another had cost $36. This made a big dent in the money he had been saving for a new bike. Stan's parents were forgiving, and, the next time they got a new contract, they included a "bucket" 1,000 text messages in the plan, where both incoming and outgoing messages count against this limit.

About a year ago, Stan was able to get a new phone, a Samsung "slide" phone. His phone has a color screen and a camera function that he uses sometimes. Some of his friends had bought the same phone and Stan thought that the camera and the color screen were cool. His phone has the ability to hook up to instant messaging and access the internet – both of these are functions that he almost never uses since they cost so much. His phone is only slightly more than a year old but, because of his active life style, it has been pretty well beaten up.

While it was a nice phone at the time, he now has set his sights on a phone with a QWERTY keyboard. He sends many text messages and he feels that the "multi-tap" system is ineffective. If it did not cost so much, he would like to try Facebook Mobile (when he gets the QWERTY phone). All his friends, and this girl he likes, are on Facebook, and some are using the mobile app. He also thinks that he could use it to update people on what life is like rafting on the river this summer.

Since Stan has only a limited income from his summer job and from a part-time job as a waiter, he cannot afford to upgrade his phone at the moment. He is still on his parents' family plan and they share 3,000 minutes a month. In addition, he still has a limit of 1,000 text messages per month.

Unlike GSM phones where the subscription is in the SIM card that can be freely moved from phone to phone, in the USA the subscription is most often inseparable from the actual mobile phone handset. This means that Stan cannot buy a new handset, but rather has to wait for his current contract to expire in about a year's time.

Stan's roommate Jerry is not so interested in mobile phones. He has a "free" no-frills flip phone he got when he signed up, and a very limited number of minutes and text messages in his contract. He does not use the phone very much and as often as not he forgets it and leaves it in the dorm room. When he remembers to take it with him, he uses the phone only when he really needs to get in touch with someone, or when he, for example, needs directions.

The coverage of the network is poor on campus and this means that Stan often misses calls that come when he is in his room. He does not have a landline phone since he already has a "cell" phone and all his friends know that number. He cannot see the point in paying for the extra phone and making all his friends learn a new number.

2 Mobile communication in contemporary society

The mobile telephone has changed the way we think of interacting. The vignettes here show how the ability to call directly to an individual affects the way we organize activities and the way that we socialize. In addition, the device allows us to interlace the remote and the co-present. In this way we can assert – or at least try to assert – a type of control over different situations. We can coordinate interaction and we can deal with various forms of emergencies, both large and small. In addition, the device itself has become a type of icon for our times. It is a way for us to show our status and to tell others who we are. Finally, the device affects the way that we integrate the intimate sphere.

2.1 Micro-coordination

Several of the vignettes illustrate the way that mobile communication has changed the way we coordinate our daily interactions. Mette needed to arrange when she should go to deliver her daughter to soccer practice and whether her husband should buy cheese or yogurt at the store. Alberto the businessman needed to arrange the logistics associated with his work, and Mika the teenager can call her mother to arrange delivery of forgotten school assignments.

The situation of Mette in particular shows the role of the mobile phone in the coordination, or perhaps micro-coordination (Ling & Yttri, 2002), of the logistics of daily life. In addition, it shows, perhaps from the perspective of the parent, the way that the mobile telephone affects the emancipation process.

With the mobile, we call directly to an individual and not just to a location in the hope that the person we wish to speak to might be nearby. This represents a breach in the way that we generally coordinate interaction. Prior to the widespread adoption of mobile communication, the logistics of coordination were based on agreements that we had entered into during our previous interaction with others. We might agree at breakfast that the husband will go shopping and that, after that, the family will meet at the movie theater for the afternoon matinee; or that after school the child will come home, do their homework and then prepare for band practice. The mobile telephone has allowed us a much finer-grained form of synchronization. It allows us more flexible forms of social coordination. It gives us the ability to interact co-presently and almost simultaneously in mediated interactions. There is less need to catch people at specific locations and times. Rather we can renegotiate our agreements in real-time and as the exigencies of everyday life make themselves apparent. Indeed, the negotiation of these interactions can be very intricate and play on various forms of tacit knowledge and expectations (Licoppe, 2007).

In many of these cases, before the rise of mobile communication, the system for coordinating our interactions with others was based on the common metric of time and the use of different timekeeping devices (Andrewes, 2002; Blaise, 2000; Landes, 1983). We agreed, for example, to meet a friend at 10.00. Using this system of interaction, both partners would independently use their timekeeping device in order to determine when they would meet up. It might be that one person's watch was ten minutes fast or slow and this would mean that that person was either unnecessarily rushed or relaxed. Regardless, the coordination of the meeting was fixed by the independent reference to a common system of timekeeping.

Mobile communication has affected the use of time-based coordination. The two individuals might still agree to meet at around 10.00 at a particular place. However, if one person is delayed, or if one of the friends is struck with the idea that it would be better to meet at another place, then this can easily be arranged by a call on the mobile. Thus, if the husband finds that tickets to the matinee are sold out, the family can decide, on the fly, to redirect their meeting to the playground in the local park or perhaps a café. Indeed, even after the wife has left home, she can be contacted and the new plans can be arranged. The point is that we have become *individually addressable* regardless of where we are and what we are up to. When trying to reach another person, we do not call to a geographical location as we do when using the landline telephone system. Rather, we call to the person.

This has led to new forms of interaction. Rather than making a relatively fixed agreement as to when and where we are to meet friends, we often engage in iterative planning. We might generally agree to meet a friend on Saturday sometime during the middle of the day. On Saturday, as the time approaches, we then go through several rounds of specifying when and where it would be good to meet. One partner might be a little delayed when the other is ahead of schedule. This might mean that the person who is first at the meeting point might have time to run a short errand while waiting for the other.

This system of interaction is best applied to small groups. When the number of individuals who are planning activities goes over some number, perhaps around eight to ten individuals, the complexities of iterative planning become too burdensome. If there is a change in the plans by one individual, all the others need to be contacted and the new arrangement needs to be worked out. In the case of larger groups, the more fixed system of time-based coordination has obvious advantages. Further, when thinking of the coordination of large institutions such as schools or airline companies, there are obvious drawbacks to the use of iterative planning. Thus, time-based planning will continue to be widely used. It is still the best system for coordinating interaction with large groups of individuals.

The discussion up to this point has shown that the mobile phone facilitates coordination and allows flexibility in what might otherwise be a stiff and unyielding time budgeting. At the same time, mobile communications facilitate "work to family spillover." This, in turn, can result in problems of increased stress and lower family satisfaction (Chesley, 2005). Indeed, according to Chesley, this may be more of a problem for women than for men. Thus, while there is a flexibility that mobile communication introduces into our interactions, this may come at the price of being too available, particularly when we are faced with the cross-purpose tensions of job and family. For many, like Alberto, Mette, and even Stan – if he can get coverage while rafting on those desert rivers this summer – the mobile phone changes the boundaries between family and work. It changes the way that we interact with friends and near family, and it also can alter the way that co-present interaction gets played out. Using the phrase of Meyrowitz (1985), it alters our "sense of place"

2.2 Health and safety

In the vignettes we saw how Stan the college student was caught out when his car stalled, how Thomas and Flores used the mobile phone when visiting family and for small emergencies, and how Mette, the working mother in Oslo, needed her phone when her mother was in the hospital. These illustrate one of the most basic functions of mobile communication –

that is, its role in helping us deal with issues of health and safety. Indeed this is often mentioned as the reason people purchased a mobile phone in the first place. Studies do suggest that mobiles help people react to emergencies, feel more secure and just stay emotionally connected over distances. Healthcare professionals use mobile technologies at every stage, from pagers and remote consultations to reminding people to take pills (Pal, 2003). Yet mobiles cause accidents while users are driving, and many people remain concerned about the health impacts of long-term mobile phone use. "Converged" features such as remote health monitoring, tsunami alerts and GPS may bring additional benefits.

If we look a bit more carefully at this issue, we see there are a variety of ways mobile use assists in ensuring our safety and security. Haddon has noted that the nature of the "emergencies" can vary. They can range from life-threatening situations to simply needing help to carry in the groceries (Haddon, 2004). Indeed, as we saw in the vignette of Stan, the young male in Los Angeles, the mobile phone can be a safety link for people who, for example, have trouble with their cars or who are often away from their safe base at home. By contrast, in the last chapter, we mentioned the very different plight of many residents in rural villages, who have no access to health care nearby, nor an ambulance to get them closer to such care. Mobiles are both a means for long-distance medical consultation and a way to call for help (Souter et al., 2005).

In extreme cases, the use of the mobile telephone in various rescue situations underscores its role as a safety link. People who become stranded or who meet unfortunate circumstances when they are in the wilderness have relied on the mobile phone in order to call for help. Some of these rescue efforts have indeed spanned the globe. People stranded on remote mountaintops in Switzerland have sent text messages to their families in the UK, who in turn have alerted the local mountain rescue personnel. People whose cars have failed in the Australian Outback, and sailors whose sailboats have become disabled, have sent messages to relatives who then helped to start rescue operations. In one example, a woman who was on a sailboat off Indonesia in the Lombok Strait when it began to take on water sent a text message stating, "Call Falmouth [UK] Coastguard, we need help, SOS," via her mobile phone to her boyfriend who was in the UK at the time. He called the Falmouth Coastguard station who in turn notified Australian and Indonesian authorities. Closing the loop, the rescue authorities called the woman on her phone so that she could help to guide them to the boat (Batista, 2001).

In addition to assisting adventurers, the mobile phone is also being used to help ensure the safety of overseas migrant workers. For example, Filipino organizations have set up systems that allow overseas workers who are being mistreated or find themselves in a threatening work situation to easily contact people who can assist them (Center for Migrant Advocacy, 2006). As noted in chapter 2, any kind of mediated communication increases security and safety. These examples illustrate how mobiles are "better" at this than landlines because of the likelihood that they will be present at the scene of an accident.

Another dimension of this issue is the notion of security, and not just in the case of dealing with accidents that have happened. As illustrated in the vignette of Thomas and Flores and in the case of Stan, the mobile phone is also used to check in with others and to alleviate the fears of family and friends. If we are unexpectedly delayed, a call will perform the courtesy of letting others know where we are and that they need not worry. Mobile communication can also be used to keep in contact with family who may need help in an acute situation, or who are, in general, more chronically in need of a secure communications channel. Palen, Salzman and Youngs (2001) for example, interviewed a wheelchair-bound man who owns a large tract of land in the USA. He kept his mobile phone with him in case he fell out of his wheelchair and became stranded. In addition, he used it when he planned to meet people, but encountered physical barriers such as stairs. A somewhat similar use of the phone is described by a Norwegian forester in a series of 1997 interviews. He noted: "When I am in the forest in the winter to harvest trees, then I always have a mobile phone in the tractor because if you are unlucky and cut yourself in the leg or something, then I am always ready to crawl over there [to the tractor] and call the necessary numbers." For people who might need assistance and for people who might come in harm's way, the mobile phone provides a link to others.

From a more general viewpoint, it is women and the elderly who are most likely to feel that the mobile phone provides them with a safety connection. Indeed, a survey of a random sample of Norwegians found that women and the elderly were significantly more likely to identify the mobile telephone as a security device (Ling, 2004). Another analysis found that women are more likely to use the mobile phone to give the impression that they are in contact with another person to "warn off" any potential attackers. More than 45 percent of the women in a small (and not random) survey of US college students said that they had pretended to use the phone when walking alone at night (Baron & Ling, 2007).[1] On a more concrete level, the mobile phone is also being seen as the repository of important contact information that can be used by emergency personnel. The use of an ICE (in case of emergency) entry in mobile phones is a way for emergency personnel to find out whom they should contact in cases where the individual is not able to give this type of information (Cohen, Lemish & Schejter, 2007).[2]

More dramatically, Cohen and Lemish have described the use of the mobile phone in the case of terrorist bombings in Israel (2005). They describe how, in the immediate aftermath of a bombing, the device has allowed families to contact one another and check on their status. The mobile phone served a similar purpose in the wake of the September 11 events in the USA (Dutton, 2003; Katz & Rice, 2003). During a period of dangerous wildfires in San Diego in 2008, there was plenty of person-to-person texting to keep tabs on the fires, but the authorities were also involved, using mobile technologies to promote public safety. The University of California system used SMS-blast messages to notify 10,000 students of class cancellations, while a local TV station used the SMS-based many-to-many stream-of-consciousness program Twitter to send updates on the fires to subscribers' mobiles and e-mail accounts.[3]

The implication is that the mobile phone has extended our ability both to reach out and ask for assistance and, in our own turn, to offer assistance to others. It provides a type of safety net that allows us to deal with both large and small issues. This has undeniable positive effects since we can more readily provide aid to one another. The cloud behind the silver lining, however, may be that we take chances where we would have been cautious prior to the adoption of the mobile phone. Blindly relying on the technology to get us out of a jam may work most of the time, but the vagaries of coverage and the fallibility of batteries might mean our link to the world fails.

At many levels, the mobile telephone helps to provide safety to individuals. The picture would not be complete, however, if it was not mentioned that the device can also reduce safety. There is no doubt that mobiles are used by criminals to organize their nefarious activities. Mobile phones have been used in the organization of terrorist activities and to explode bombs by remote control, and yet have also been used to expose terrorists in hiding.[4] There is an ongoing debate and inquiry into the existence or strength of a relationship between mobile telephones (and mobile towers) and physical health; some feel that the radio waves emitted by the devices and their towers contribute to cancer and other ailments (Burgess, 2004).

Perhaps the clearest threat they pose, however, is in their relationship to driving. Virtually all research on the topic shows that talking on the mobile phone while driving increases the risk of accidents. The data from the study of actual accidents (Cain & Burris, 1999; Redelmeier & Tibshirani, 1997), as well as studies that have examined the interaction between mobile phones, driving and auditory disturbances (ICBC, 2001), talking into the device (Strayer & Johnston, 2001), visual aspects (Strayer, Drews & Johnston, 2003) and different combinations of these disturbances (Burns et al., 2002; Recarte & Nunes, 2000), all agree: talking on the phone while driving is dangerous. This is particularly true for younger drivers (Strayer & Drews, 2004). Indeed, in laboratory comparisons, talking on the mobile phone introduces impairments as profound as those seen in drunk driving (Strayer, Drews & Crouch, 2006).

2.3 Mobile communication and the display of self

In addition to issues of coordination and safety, people use mobile phones for reasons that cannot be easily described in such instrumental terms. While some people are more or less indifferent to the design of their mobile phones – take, for example, Stan's roommate – other people focus a lot on the style and vintage of their devices. In the vignettes, the businessman Alberto was very intent on having an advanced telephone with many functions. While his use of the phone was instrumental, the actual form of the telephone was also quite important to him. In a similar way, the handsets of Mika the teen and Stan the college student were also visual cues that other people used when determining what type of person they were. The mobile phone that we carry with us is an element in our display of self (Fortunati, 2005b).

Mobile phones have been status symbols from the start.[5] They show no sign of losing their appeal, as new models arrive virtually every week. Research suggests that there are sociological and psychological explanations for these display behaviors – the mobile is a cultural object/artifact, with important display and signaling features. We can have a business mobile phone, a nerd mobile phone, a high-fashion phone, a feminine or a masculine mobile phone. We can have a phone with European flair, Japanese functionality, Korean design or American ingenuity. The phone can be high-tech or a throwback to a previous era. The point is that we actively interpret the phones of others and we are careful in the selection of our own device. At one level, when we are shopping for a phone, we think about its functionality. At another level, however, we are concerned with the design of the device and how it is an element in our display of self (Fortunati, 2005c). What image do we want to project? In addition, we often glean clues about others' sense of themselves by observing which type of mobile phone they use. The mobile phone has become one dimension of how we construct our own identity. The mobile phone is not just a functional item; it is also a symbolic item. The selection of mobile telephones and mobile telephone services is not only a signal to others of how we wish to be seen, but also a way that we integrate our self-image (Pedersen, 2005).

Thinking somewhat more broadly, fashion is a slippery concept. What is "in" today will be "out" in a short time. In addition, the fashion style of one group stands little chance of being adopted by other groups. Indeed, the nuances between groups can be quite fine-grained (Lynne, 2000).

One of the central analyses of fashion was carried out by the German sociologist Georg Simmel (1971). According to Simmel, fashion consists of two dimensions, or perhaps two tensions. The one dimension is the tension between being *avant-garde* and dowdy. The other is the tension between using fashion to indicate our individuality vs. the use of fashion to show group membership. Timeliness is an essential issue with fashion. Indeed, being fashionable is a type of sweet spot between being ahead of our time and being dated.

Those who, for example, paint their fingernails black, who choose seemingly exotic colors for their clothes, and who have the most bizarre ringtone on their mobile phone are those who push the boundaries of consensus. At the other end of this spectrum are those dreary individuals who wait too long to adopt a style and are perpetually a couple of steps behind the broad consensus. Being in fashion is obviously a relative term. We can be "soooo last month" or "soooo last century." The fashion cycle can move very quickly – or tragically slowly – for some people and for some elements in their display.

There are those who note that the style of lapels or the color of scarves changes with the seasons. There are also the archetypical maiden aunts or corny uncles who resolutely maintain the style of a previous generation. In the middle of all this are those who somehow seem to balance between being too advanced and too dull. Having a sense of this tension is one of the dimensions of fashion.

The other dimension of fashion is the balancing between individual expression and group identification. This is perhaps illustrated by Mika and her selection of a mobile phone. On the one hand, we often see fashion as a way to indicate to others who we are. There is a drive towards individual display and, at the same time, towards group identification. We want others to admire the qualities of our taste and flair. At the same time, we do this using a range of expression that is common to others in our reference group. We choose a jacket, a pair of boots, a necklace or a mobile phone that we feel expresses our own sensibility. However, the choice of jackets, boots, necklaces or mobile phones from which we choose should be within a certain range. Thus we see: "Two social tendencies are essential to the establishment of fashion, namely, the need of union on the one hand and the need of isolation on the other" (Simmel, 1971, 301).

Our expression of fashion is also the establishment of a boundary between in-group and out-group. Indeed, fashion is an active way through which we express our allegiance to a particular clique or social grouping:

> From the fact that fashion as such can never be generally in vogue, the individual derives the satisfaction of knowing that as adopted by him it still represents something special and striking, while at the same time he feels inwardly supported by a set of persons who are striving for the same thing, not as in the case of other social satisfactions, by a set actually doing the same thing. (Simmel, 1971, 304)

The mobile phone fits into these two dimensions just as do other personal artifacts. While the mobile phone is a device used for instrumental activities ("When are you coming home?,"

"Where are you now?," etc.), it is also very clearly a fashion item. In this regard, it is to some degree a symbol of group association. New phones have functionality that makes older models seem clunky and out of touch. A lot of fuss can be made in association with the launching of a particularly new type of handset. The new phone might have some functionality that is not contained in current models. It might be able to download things faster or have a different set of buttons, controls or touch screen. At the same time, the motivation for buying it might simply be that it has a sleeker profile.

Given that the mobile phone is a fashion item (in addition to being a lot of other things), this means that it is an item about which we need more or less constantly to be aware of what is coming (Fortunati, 2005c). To be an aware mobile user, we seemingly need to know if the new thing is 3G, has an MP3 player, a touch screen with new functionality, a camera with better resolution or a new way of connecting. This also means that we collect older phones that in their time were the objects of our lust and desire but have become curiosities – and risk becoming environmentally damaging cyber junk if not re-sold (and possibly made fashionable again) in the second-hand markets.

The style of mobile phone handset we have tells others a bit about who we are and how we want to be seen. Indeed, some people have several different handsets that they use according to the situation. They may have an everyday "business handset" and another one they use when they go out for the evening.[6]

2.4 The dynamics of family integration

Mobile use, as well as landline telephony, changes family dynamics. Both forms of telephony allow for interaction with distant persons. However, it is the mobile that increases the potential for mobility and allows mobile for, for example, Rakow's "remote mothering" (Rakow & Navarro, 1993). Using the mobile phone is a way for families to communicate amongst themselves. As with Mette, the professional working mother in Oslo, the device allows for internal coordination and the arrangement of practical logistics. It is in the flux of family interaction that we see the importance of personal address-ability, the interlacing of messages into the folds of daily life and the use of the technology to at least try and grasp a certain sense of control over our everyday affairs. In addition, the mobile increases the freedom of teenagers.[7] It is not an exaggeration to speak of the mobile phone as an element in the emancipation of the teenagers of the world.

According to Chesley (2005), the use of the mobile telephone in the family context is not an unmixed blessing. In her analysis of people who were largely well educated and in white-collar jobs, she found that use of the mobile telephone increased the "spillover" between work and family. As was illustrated in the vignette of Mette, the device allows for issues within the family to intrude into working life. In addition, there is the opposite form of spillover – that is, their jobs intruded into their private lives. Chesley has found that this type of spillover has negative effects on family satisfaction, and particularly for women.

The use of the mobile phone by divorced parents also shows how we use it for both practical and expressive interaction. Norwegian research has shown that children of divorced parents often get a mobile telephone somewhat earlier than their same-age cohorts (Hjorthol, Jakobsen, Ling & Nordbakke, 2007). There are many practical issues to be dealt with when children live in two separate homes. Children in these families have to keep track of various items (favorite stuffed animals, sports equipment, schoolbooks, etc.) and the mobile telephone helps in the arrangement of delivering children to the right place at the right time. In addition, the mobile phone provides a direct link between the child and the non-resident parent. There is no need to call the landline phone of one's "ex" with the possibility of a potentially difficult confrontation. Rather, the non-resident parent can call directly to the child.

Thus, we are still in the process of working out our relation-ship to the mobile phone. According to the ideas forwarded by the so-called "domestication approach," the device has not found its final niche in our lives (Haddon, 2003). It was bought with one idea in mind and then it has developed and changed as new functionality has been added and as users have found how to apply this functionality in their daily affairs. In addition, there is a temporal aspect to domestication. For example, Mette only used the mobile phone in a limited way when she first received the device. However, as time has gone on, she has come to use it more often, and her use has grown to include employing it in many different situations.

3 Conclusion

Taken together, the ten vignettes presented in these chapters illustrate how widespread and varied mobile use has become over the first two decades of major use. One interesting theme in the later set of vignettes is how similar some of the uses are to the uses in the vignettes presented in chapter 3. Rohit, Mette, Alberto and Annette use their handsets to micro-coordinate – Annette guides her customers to arrive at the right time for a hot lunch, Mette keeps control of the different family members, Rohit guides his pals to the right café for a hot coffee. Safety, too, is on the minds of many "first-time" as well as experienced mobile users. For example, Liang used his new mobile to get valuable information (and to feel connected) during the SARS crisis. Annette, Mika, Liang, Stan and Rohit have all chosen "fancier" handsets rather than the least expensive ones available to them. Their tastes are different, but their motivations are similar; each wants to use the mobile to "show off" a little bit, to say things about themselves to their friends and neighbors. In many ways, using a mobile telephone in China or Bangladesh or Nigeria is no different from using a mobile telephone anywhere else; people make and receive voice calls, compose text messages, micro-coordinate, check in for health and safety, and show off their new purchases. Mobiles are used by rich and by poor, by young and by old; mobiles are used for purposes saintly and nefarious, and for the frivolous as well as the monumental. Mobile use is truly worldwide.

Mobile phones have captured the day. Their rate of diffusion is among the fastest of any communications device; indeed, it is rare to find teens in developed countries that do not have one. We adopt them because they provide us with the ability to communicate with friends and family. They are relatively inexpensive to use and they have a type of fashion currency. The handsets – in their many forms – are stylish and the tariffs are not an insurmountable burden. The fact that we are using them to call and to send literally trillions of text messages says that they have fit neatly into a niche in our social ecology. Many people are passionate users of a steady stream of advanced services – the mobile internet, global positioning systems, electronic payment, mobile TV, etc. – and are early indicators of social changes still to come. However, the basic notion of whenever/wherever communication is the true revolution already accomplished by the mobile telephone. The rest of the stuff is icing on the cake. The fact that we can be in touch with our intimate sphere is the motivating force behind the adoption and use of the mobile telephone.

At the same time, mobiles are not landlines – they do different things, and threaten/promise to bring about 21st-century, post-modern forms of interaction. Mobiles are not just landline substitutes – they can do things and go places where landlines cannot. They add dimensions of flexibility into our daily schedules and they often allow us to compress activities. At the same time, these characteristics mean that traditional boundaries between co-present and mediated interaction become difficult to pin down. Though we think of "road warriors" like Alberto in Milan taking advantage of the mobile's portability to "be reachable" and to do business wherever he likes, these same properties are helpful in other settings. Many people conduct their livelihoods roaming from place to place or without a fixed address. Mobiles help taxi drivers, delivery people, plumbers, etc., be reachable anytime, speeding the matching of buyers and sellers and accelerating the "the urban metabolism" (Townsend, 2000). Usually carried by people rather than bolted to a wall or resting on a desk, mobiles make people more reachable, regardless of location, and thus alter our relationship with space. We are individually reachable and addressable. It does not matter if we are en route or simply not immediately at hand. We are available to potential interlocutors through the simple dialing of a number, just as they are available to us. It is true that the mobile phone is not the first technology through which we are able to communicate with remote partners. The telegraph and the traditional landline telephone have both afforded this type of connectivity. The mobile phone, however, individualizes communication. Thus, instead of calling to a central office or a home, we call to an individual.

A host of features, particularly text messaging, morph and extend the "traditional" voice call into asynchronous forms of communication, altering our relationship with time, and the mobile's remarkable processing power (both on the handset and over the network) is bringing data, images, music and videos to our fingertips; as we become closer to our "data," our memories, our work and our surveillance of our social environment are mediated by our handsets, and our mobiles alter our experience of mind.

In the next two chapters, we will consider some of these uses through various theoretical lenses, focusing first on some of the debates brought about by widespread mobile use, and later returning to some of the theoretical discussions of the mobile's place in the information society.

CHAPTER FIVE

Debates surrounding mobile communication

1 Introduction

The mobile has not snuck into our lives. Rather, as we can see in the vignettes in chapters 3 and 4, it has come crashing in – welcomed and banned, debated and celebrated, scorned and venerated. The mobile is a symbol of the modern and the post-modern, of individual autonomy and social connectedness, of independence and collective action. Few technologies have captured the popular imagination like the mobile phone (cars, computers, trains, light bulbs and airplanes come to mind). Countries around the world celebrate unique mobile phone cultures and habits, and yet share the same technologies. In this chapter, we will examine several issues, often debated in the popular press, with regards to mobile communication. These include issues such as disruption of the public (and private) sphere, surveillance and the inappropriate distribution of revealing photos, the destruction or at least reframing of the language and various types of power disruptions at the personal and at the social levels.

2 Disruption of co-present interaction

While giving a speech during the 2008 US presidential primaries, candidate Rudy Giuliani was interrupted by a call on his mobile phone. He was talking at the conference of the National Rifle Association – a supremely important lobby group for a Republican candidate – when his wife called.[1] Upon hearing his phone ring, he interrupted his speech by saying: "Let's see now. This is my wife calling, I think." He took the call and said:

"Hello dear. I'm talking, I'm talking to the members of the NRA right now. Would you like to say hello? [pause] I love you – and I'll give you a call as soon as I'm finished. OK? [pause] OK, have a safe trip. Bye bye. [pause] Talk to you later, dear. I love you." Before proceeding with his prepared remarks, he noted, "Well this is one of the great blessings of the modern age, being always available. Or maybe it isn't, I'm not sure."

The episode, and Giuliani's ambivalence, summarize our attitude towards the use of the mobile phone in the public sphere. As we have noted, the mobile telephone provides us with whenever/wherever access to friends and family. While this is a convenience that facilitates interaction, it also means that the mobile phone can spontaneously announce its presence at the most unexpected times and in the most awkward places, disrupting the norm of co-present interaction. Conversations about these awkward moments are as old as mediated communication – people grumble when a landline rings during dinner – but have taken on more urgency as mobiles have expanded the opportunities and locales for possible disruptions.

The fact that Giuliani received a call while he was giving a speech is an ironic twist on the more standard situation where performers – and the other patrons – are disturbed by a ringing phone in the audience, not on the stage. For example, actor Richard Griffiths – the man who plays Uncle Vernon in the Harry Potter films – interrupted a stage performance in a London theater upon hearing a mobile phone ring in the audience. Using his best bluster, he said:

> Okay, I am not going to compete with these electronic devices. You were told to turn them off by the stage manager, you were told it was against the law, and you heard two phones go off already before this. You should be ashamed of yourself. Now I'm going to exit, and we're going to start this scene again, so tech stand by . . . and I assure you if we hear one more phone go off we'll be in our right mind to quit this afternoon's performance . . . you have been warned.[2]

Griffiths' sentiment is not lost on teens, who are not normally among those who are the most accepting of mobile phone use in public places. Indeed, in a group interview carried out in 2000 in Norway, the topic of telephone use in public places was discussed:

Moderator: Another problem that comes up is that [the mobile phone] disturbs public places.

Grethe (19): In the movies. That is really irritating.

Moderator: Have you heard this lately?

Gro (18): I was at the movies the other day and I heard it. It was just awful. They know that they need to turn off [their phones] at the movies.

The transcript shows that there was a well-understood ethic among the teens with regards to the use of mobile phones in movie theaters. While there were breaches – which were perhaps sins of omission rather than sins of commission – the underlying attitude is the same as that expressed by Griffiths.

On a slightly smaller stage, the use of the mobile phone in meetings raises some complex issues. There is not the same large assembly and often there is a multisided interaction in the meeting. Nonetheless, a call on the mobile phone can interrupt the flow of the interaction. Interestingly, it can also expose the status and power dimensions that are at play. There is an ill-defined challenge to etiquette posed by the device. Who can accept a call while at a meeting? Is it OK for a boss, but not underlings? How does texting fit into the mix (Leland, 2005)? In general, those in high-status positions have greater prerogative in their use of the mobile phone.[3]

Giuliani's phone call and the call that raised the ire of Griffiths were both disruptions played out in large format. While they were spectacular, they were not the most common form of interruption associated with the mobile phone. In addition to disturbing larger public gatherings, the mobile phone can interrupt smaller-scale forms of interpersonal interaction. All of us have listened to the boorish person talking on his/her phone in a very public place. Some people go so far as to (illegally) get a jamming device just to protect themselves from these situations. These stories, regardless of their validity, speak to people's unease regarding public use of the mobile phone and the feeling by some that it needs to be restricted and controlled (Richtel, 2007).

All of us have had to put our face-to-face conversation on hold while our interlocutor answered their phone and, if we are completely honest, most of us have done the same to others. Mobile phone conversations in public were unknown in the recent past.[4] Research shows that it is younger users who are more tolerant of mobile phone use in public places – though, as noted above, they also have their sense of when and where it is not appropriate to use the mobile phone. While a survey of 752 people in the USA found that about 6 out of 10 individuals think of public use of mobiles as "a major irritation," a majority of younger users, aged 18–27, disagreed (Traugott, Joo, Ling & Qian, 2006).

Use of the mobile phone in public is a convergence of the public and the private selves. Goffman, who studied non-mediated interaction, noted that there is a need to manage the distinction between the front and the back stage (1959). For him, our public "front stage" activities involved the management of a display. When in public, we are aware of how we present ourselves, our deportment and our façade. We are often minutely concerned with what he calls "impression management." By contrast, the back stage is where we can let the façade slip and we can engage in the more mundane maintenance of our image. We can check our hair, adjust our clothing and we can agree with others as to how to deal with a new version of the public performance. The mobile phone disturbs this bifurcation. As can be seen in Giuliani's telephone call, the mobile phone conversation – including the seemingly mundane coordination with his wife and the use of endearments that are not often heard from political candidates as they are delivering a major policy speech – meant that back-stage activities were quite literally pushing themselves onto the front stage. Activities that we might otherwise carry out in more unguarded moments were thrust into co-present, front-stage situations.

Given the fact that our telephonic interlocutors are not cued into our current physical situation, there is always the possibility that the phone call will spin out of control. The tearful and loud break-up of a romantic relationship, the maneuverings of a back-room deal or the exercising of corridor politics can be done telephonically, whenever, wherever and with unanticipated audiences on both ends of the line. Thus, the co-present audience might get to know more than we would wish. This is further complicated by the fact that the co-present situation can also thrust itself into the mediated interaction. It might well be that the telephonic interlocutor assumes that the conversation is "just between us two." However, when they hear the tinkling of glasses and the young man or woman giggling in the background, they might suddenly recognize that the discussion is not as closed and guarded as they had originally thought.

The staging of a mobile phone conversation is fraught with problems. To use it successfully in public requires a certain *savoir-faire*. The ringing of the mobile phone interrupts the flow of a co-present situation. With the speech of Giuliani, the performance of Griffiths and the movie experience of Grethe and Gro, there was interruption of a pre-existing situation. Similarly, in the case of a conversation between friends, the call on the mobile phone also interrupts the flow of interaction. In a conversation, we exchange comments and rejoinders and follow certain topics with our conversation partner(s). The interaction might be focused on how to deal with a certain problem. It might be a "pitch," where one of the interlocutors is trying to convince others about something, or it might be simple banter between friends. Regardless, the ringing of a mobile phone disrupts the flow.

We can be more or less adroit at dealing with this. Indeed, it is through the nimble use of manners and courtesy that we facilitate the flow of events and buffer them from unwanted disturbances. Further, when a disturbance has indeed happened, it is our sense of courtesy that helps us to negotiate through the shoals of social interaction. There is the role of the person responsible for the disturbance and there is the role of the individual who must suffer the disruption in the flow. Each partner has a range of devices that he/she can use. The person receiving the call can ignore it, they can feign resignation as they answer, they can apologize to the co-present interlocutor as they transfer their attention to the mediated interaction or they can simply stop the pre-existing interaction and take the call. In the same way, the "injured" party can bluster at the interruption, they can nod discreetly and permit it or they can try to continue the interaction in spite of the emerging situation.

All of this changes, of course, if one of the co-present individuals needs to initiate a call. Rather than being an unanticipated breach in the interaction, there will likely be some sort of negotiation before the actual call is made. The point is that the individual initiating the call will have to mark their withdrawal from the co-present interaction in order to make the call. In that situation they can use forms of courtesy, manners and deportment (Goffman, 1959). After more than a decade, the use of the mobile phone in public places is still an unsettled issue.

3 Mobile communication and power relations

One theme that arises in this book is that the mobile phone is mostly used for mundane interactions. It is used to coordinate who will meet whom, when and where. It is used to decide whether he or she will make dinner tonight and it is used to arrange the ten-minute delay of a meeting because traffic is particularly heavy. Many, if not most, of our calls are like this. The vignettes in chapters 3 and 4 describe this type of interaction.

Of course, one result of all these little interactions, most of which pass and accumulate without any of us really thinking too much about them, is a slow layering-on of the networks and

habits of our social lives; these have an implicit power component. Who calls whom with the reminder about picking up
dinner is both a reflection and a reinforcement of the matrix
of expectations and obligations occurring in a relationship or a
family. When an employee sends an intentional missed call to
his employer, asking the employer to pay for the call (because
the social norm is that richer people have to pay for calls), he
is acting within, acknowledging and ultimately strengthening the power relationship between employer and employee
(Donner, 2007).

Yet beyond these everyday interactions, there are occasional
situations where we are able to see the way that the mobile
phone affects power relations in a much more overt way. While
it is possible to say that there are power dimensions associated
with our everyday calls, it is also possible to discuss power in a
much broader sense, i.e. the situations that actively challenge
social norms and structures, and raise "debates" in the popular
discourse. A good definition of power has been provided by
the sociologist Max Weber, who says that power is when one
person or a group of people is able to realize their will "against
the resistance of others who are participating in the action"
(1958, 180). This can be legitimized through traditional, legal–
rational or charismatic structures. In several cases, mobile
phones have been involved in the shifting of state power,
they have been the instruments of terror and they have been
employed by the authorities to capture the bad guys. At a lower
level of abstraction, mobile communication has democratized
and complicated our ideas of news-gathering, disempowering
traditional arbiters of news, and it has affected power relations
within the family. In this section, we will examine the way that
mobile communication affects power relations in both large-
and not so large-scale interactions.

3.1 Mobile communication and social protest

One of the unexpected consequences of the mobile phone is
that it has, in some cases, resulted in a reallocation of political

power. There are political mobilizations that are organized by mobile telephone, as well as terrorist activities. In some cases, the boundary between the two is difficult to determine. One person's political engagement is seen by another person as subversive. Regardless of the perspective, issues of power are in play and mobile communication has become a tool in these interactions.

One of the first, and indeed the most celebrated, examples of this is the ouster of President Joseph Estrada in 2001 in the Philippines. In that case, thousands of protesters collected at a place in Central Manila called the Epifanio de los Santos Avenue, known in daily life as the EDSA. They were alerted to the protest with the simple SMS message "EDSA." The location was not without precedent since it was also where there had been earlier protests against Fernando Marcos. In the case of the Marcos protest, the military, who are housed at a nearby army camp, refused to break up the protest and in effect ended the reign of Marcos. Thus, the location is symbolically imbued in Filipino history.

It is of course obvious that the actual protests against Estrada arose out of a much broader political situation. It is important to note that the mobile telephone did not cause the protests, but rather its use aided in their organization. Estrada, a former film star, was elected by capitalizing on his links to the more impoverished population. At the same time he was deeply unpopular among some portions of the populace, notably those who were better able to afford items such as mobile phones (Rafael, 2003). He had been accused of corruption and was the subject of an impeachment process that became the focus of national attention. The impeachment process came to an impasse involving the decision to introduce a particular piece of evidence into the trial. The impasse resulted in half of the panel that was considering his impeachment – those who were opposed to Estrada – walking out. It was at this point that the protesters started to send out the message "EDSA" to all the individuals on their mobile phone lists, indicating that people should congregate in order to voice their opposition (Paragas, 2003).

Taking this somewhat broader perspective, it is easy to see that texting did not cause the protest, rather texting facilitated it (Pertierra et al., 2002). Thus, there was a generally recognized tension (the assertions regarding Estrada's corruption) and a widely shared ideology. There was a dramatic turn in the impeachment (the impasse) and there was a well-recognized, symbolically imbued location that could be employed for the protests (EDSA). Finally, there was an easily used form of communication that allowed for the quick mobilization of the protesters. Conveniently, the message "EDSA" could be quickly sent out via SMS. Thus, the SMS being sent out was both a form of group coordination and also a call to arms (Ling, 2004). It is important to note that all the elements were important and, were one of them to be missing, the protest would have fallen flat or taken a different form.[5]

The anti-Estrada action in the Philippines, the various World Trade Organization protests, the protests at the 2004 Republican Convention in New York, the election of Viktor Yushchenko in 2004 in the Ukraine, and an ever-increasing number of examples point to the fact that the mobile telephone has become a part of the political landscape. It is important to note, however, that, just as the mobile phone can facilitate the logistics of social protest, it can also facilitate the tendencies that would minimize the impact of social protest. Writing before the rise of mobile communication, William Gamson wrote an interesting book on the strategies of social protest. He examined the histories of 53 protest movements – of all types – in order to better understand the dynamics of those that succeeded and those that were either co-opted or defeated (Gamson, 1975). A key characteristic of those groups that were defeated was that there was a tendency towards factionalization (see also Frey, Dietz & Kalof, 1992). Those protest groups that had the ability to maintain a unified focus and ideology were the most successful. Those that were plagued by competition between different factions were not as successful in carrying out their agenda. Carrying this thought into the realm of personalized mobile communication, it is clear that, while the mobile telephone can help to facilitate the logistics of protest, it can also facilitate the logistics of ideological splits. Just as we can receive a message that the protest will be held on a certain day at a certain time, we can receive a message from our more immediate group of comrades that our participation will be conditional or that we will participate in a particular way.

Elements of this were seen in the protests associated with the 2004 Republican National Convention in New York. In that case, mobile communication was used extensively to mobilize protesters for a wide variety of events and actions. There was no lack of creativity in the protests that were staged. However, at the end of the day, the protests did not significantly disrupt the actual Republican Convention. Indeed, there was not a single unifying focus to the protests. There were people opposed to environmental policies, to the war in Iraq and to a variety of other causes. Each of these was given the room to play out their own special protest at their own special time in their own special location. If we compare the impact of these protests with those in Manila where a government was overthrown, it is possible to see the difference in magnitude. In the case of the Estrada demonstrations, there was a broad-based movement that was unified in purpose and had been called to action in a dramatic national moment. In the case of the protests at the Republican National Convention, if we are to be a little flippant, we saw a type of demonstration that was more characterized by a kind of "theater of the moment" than in a true political movement. The mobile phone facilitated the protests both in the Philippines and in the USA. In the latter case, however, there was not the intensely felt focus on a single clear goal that characterized the former.

There have been several cases where protesters have used mobile communication to organize themselves. More recently, we have seen the authorities becoming wise to this possibility and taking measures to counteract it. In the 2006 protests in Nepal and in 2007's pro-democracy protests in Myanmar (Burma), the government simply turned off the mobile phone system.[6] This was done in order to stifle protests and hinder the ability of protesters to organize.[7]

3.2 Everyday citizen empowerment

In addition to affecting the machinations of power in dramatic, but focused, protests, the mobile phone is also an element in the way that power is worked out in more mundane situations. One example is the oddly named "virtual community" associated with Trapster.[8] It encourages members to send in the locations of "live" police speed traps via their mobile phones, including police cars and motorcycles that are "hiding in bushes along a freeway on-ramp" and various types of hiding places used by police and for remote sensing devices intended to capture speed violations. Interestingly, the promotional material for the service evokes democratic ideals by paraphrasing Lincoln's Gettysburg Address. It asserts that it was "created by the people, for the people," presumably asserting that it is a natural right to exceed posted speed limits.

There are also initiatives designed to use the technology for making the streets safer. One example is the initiative developed by the New York Police Chief's Benevolent Association designed to help in reporting drunken drivers.[9] The material for the campaign describes the characteristics of drunken drivers – weaving in and out of traffic, making sudden abrupt stops or starts, etc. – and it asks drivers to call the emergency number to report these individuals. It even goes as far as offering a reward of up to $250 for making these reports. Another more vigilante type of approach is seen on sites such as RottenDriver.com, where people send in a text message with the license plate of poor drivers. These are then tabulated into scoreboards for different locations in a type of public shaming. Similar traffic reporting programs have been developed in Malaysia and in the UK. In addition, mobile phone-based reporting is also being applied to pollution violations in Jakarta, vandals in Edinburgh and disease outbreaks in Indonesia and China.

These different programs do not use the mobile phone in a fight for the future form of government, such as we saw in the instances of the Philippines or Ukraine. There are nonetheless issues of power being worked out here. On the one hand, people who resent the use of speed traps are using mobile communication to organize themselves in order to assert their will. They are developing and using systems that help them to out-fox the police. At the same time, there are those who use mobile communication to help reduce the risk from errant drivers, vandals and other public nuisances. The total impact is difficult to gauge. It is difficult to know whether the technology is facilitating more speeding, more traffic tickets or both. There is, however, responsiveness in the system that was not there before.

Taking a somewhat broader view, information is being exchanged and gathered for these various and sometimes contradictory purposes. We are also seeing the engagement of people in the public sphere via their mobile phones. There is a type of participatory culture. It can range from reporting speeders to reminding others to vote, and in some cases to encouraging others to participate in a protest.

3.3 Citizen journalism
Mobile communication is also changing power relationships by facilitating citizen journalism. It makes possible the spread of information on celebrities and enables non-professional journalists and paparazzi to make in-roads into the production of news. In other cases, mobile phones (*qua* cameras) are being used to record breaking events, for example Saddam Hussein's execution, which was covertly filmed on a cell phone and later uploaded to the web.[10] The fact that these devices are becoming ubiquitous, and indeed a part of the "blogosphere," means that they are available when other recording devices are not, or when the unforeseen takes place. That is, they are

available to capture newsworthy events and these photos are finding their way into the news (Koskinen, 2004). Recent advances are cutting out the "upload" step altogether. Many phones and applications allow for immediate uploading of photos to web sites, social networking platforms and photo-sharing sites. The same is now occurring with video shorts on mobiles; third-generation networks are fast enough to support real-time streaming from camera phones to the web, allowing almost anyone with a camera phone to be a participant in the creation of "news." Indeed, CNN and other sites welcome such content from viewers. Whether these citizens are "journalists" remains a matter of some debate and is largely a matter of context, content and one's ideas about what constitutes journalism. Posting a video of, say, an arrest made at a political protest to one's own Facebook page may not be journalism, posting it to your blog might be journalism (depending on who reads your blog and why), and posting the same video to the CNN web site almost certainly is. This is making the job of being an editor more complex (Baldwin, 1997; Sonenshine, 1997). Does, for example, reporting on a politician's major policy statement on the economy serve the public interest better than reporting on his or her spontaneous racist rant that was captured with an unnoticed mobile phone? The photos of guards abusing prisoners in the Abu Ghraib prison and the video of Saddam Hussein's execution are being drawn into the internet-based blogs and thus into the broader discussion of current events. As seen in these occurrences, it is clear that the mobile phone is part of a "citizen" or "participatory" journalism movement, which is sometimes more direct, more open to public comment than traditional forms, and is providing the news consumer with alternative takes on events (Allen, 2007; Lasica, 2003). In chapter 1, we described Jenkins' description of an emerging "convergence culture" (Jenkins, 2006), in which the roles of producers and consumers across formerly distinct media platforms are no longer clear and are increasingly intertwined. By challenging the role of newspaper editors and TV producers and professional photographers as the arbiters of the news, mobile-enabled citizen and participatory journalism is a clear reflection of this convergence culture.

The ability to bypass editors is particularly convenient when you have something you want to share with a small group of like-minded people. Rather than relying on professional news-gathering organizations, interested individuals can gather and broadcast news that is of particular interest to a smaller group in society. Through internet-based systems such as UPOC (Universal Point of Contact) or the arrestingly named Gawker Stalker, group members can contribute texts on various themes (Rheingold, 2003). These are then broadcast to all other members, who can view them either via a PC or via their mobile phone handsets. One of the practical uses of this – and other similar systems – is its use in spreading "celebrity sightings." From these locations, we can follow the movements of various well-known individuals. A recent log on New York City "celeb sightings" included the following entries.

> Oct 04 2:00PM From: PA – Pharell spotted outside the Hammerstein Ballroom on W34th st in Manhattan
> Oct 05 10:04AM From: AD – Tom Cruise in Macy's on 34th Street
> Oct 05 5:13PM From: AD – Robert Dinero in Starbucks at One Penn Plaza on 34th Street bet 7th and 8th
> Oct 05 5:15PM From: FA – cool
> Oct 08 10:09AM From: ER – Myley Cyrus and Billy Ray Cyrus on East 67th between 2nd and 3rd avenues in NYC at 10:05 AM
> Oct 09 4:37PM From: CM – tony bennett in town in NBC area

The slap-dash form of writing underscores the authors' capacities with texting as well as the absence of editorial intrusion. Lower-case initials for names and various misspellings – Myley vs. Miley, Dinero vs. De Niro, etc. – tell the reader that this is breaking news. In addition, there is often a skewed distribution of contributors vs. readers.[11] Regardless of this, the entire "news cycle" is being carried out in quasi-real time and it is being done without the intervention of traditional professional groups.

Access to mobile communications devices allows for new channels in the collection of newsworthy – and questionably newsworthy – information. This changes the role of the individual vis-à-vis the news-gathering system. It challenges notions of the professional photographer as the arbiter of visual content. It challenges the role of the editor as the arbiter of which items will be distributed to the public and, in the case of celebrity journalism, it challenges the inviolability of well-recognized persons as they move about in society. The image has its own set of assumptions and is, in its own way, a source of social disruption. Things that capture images are disruptive in a different way from communications devices. People are using their mobile phones to capture everyday events.

3.4 Interpersonal forms of power relations

The mobile telephone – sometimes uncomfortably – plays into the interaction between parent and child. This interaction can be read using the lens of power relations (Ling & Yttri, 2006). Mobile communication has come to play a part in the emancipation process of adolescents (Ling, 2009). On the surface, it provides a convenient way to maintain contact and coordinate interaction. At the same time, however, its use can be shaped and sabotaged in different ways to the advantage of children in one turn of the interaction and to the advantage of parents in the next. It can be used to control information and it can be used to make – and deny – commands.

This also affects interactions between employers and employees. In her analysis of live-in maids in Singapore – like Amihan in chapter 3 – Sun (2006) examines the way that use of the mobile phone by maids is discouraged by employers. At the same time, the mobile is valued by the maids as a way to maintain contact with family and friends. In addition, it facilitates their ability to compare working conditions with other maids. According to Sun, employers were interested in guarding against this type of information pooling: "When maids get together, they tend to compare their jobs . . . salary, sleeping hours . . . when they see differences, they are not happy. That's why I get a maid with no off-days then she can't go out to mix with other maids" (Sun, 2006, 6).

Employers also limit the maids' access to traditional landline telephones and discourage social interaction. This form of limitation also extends to mobile phones, or "hand phones" as they are referred to in Singapore: "I do not feel that [my maid] should carry handphone as they can make other contacts, do illegal things or get unnecessary influences. My maid does not carry a handphone" (Sun, 2006).

As we might expect, the maids have a different perspective. Ownership and use of a mobile phone allows them a certain privacy, it allows them to contact family as well as friends (both overseas and locally) and it allows them the ability to contact authorities should that need arise. This can be either the agency they use to find work or, in some cases, embassy personnel. This does not mean that it is taken for granted that these domestic workers are allowed to have mobile phones. Sun documents how the maids need to hide them and use them surreptitiously.

Thus, the adoption and use of the mobile phone has resulted in a discussion of how power relations are exercised in different settings. Coming back to Max Weber and his discussion of power, mobile communication has influenced the way that we realize our will when confronted with the resistance of others. This has been done in both large- and small-scale interactions, to the joy of some and the chagrin of others. In some cases, lone individuals or small groups use the mobile phone to enhance their position relative to others (such as the case of the maids vis-à-vis their employers). In other cases, the mobile phone is a device that leaves a "digital trail" of information that others can follow and, in some cases, use in order to assert their power over the user.

3.5 State power and surveillance

One final issue regarding the interaction of power and mobile communication is the use of mobile phone traffic information for various types of surveillance. The traffic data accumulated by mobile phone operators can describe the web of calls that an individual makes and receives (González, Hidalgo & Barabási, 2008). This, in turn, can be used to understand the way that persons and groups with nefarious intentions communicate and operate. Nefarious activities are, of course, defined in the eyes of the beholder. They can be the plottings of small-scale criminals or other desperados, they can involve terrorism against a legitimate government or they can be the activities of well-intentioned conspirators plotting against a despotic dictator. Those who control the telephone network, regardless of their moral trajectory, have the potential to use the information that is generated by the system to follow the movements of particular handsets – and presumably their owners – and to understand the structure of social networks.

The Bush administration in the USA, for example, was criticized for its use of telephone records to trace the interactions of people who are suspected of terrorism and to map out their social networks. Civil rights campaigners have reacted to this by saying that it is an unnecessary invasion of privacy. In addition, since the actual content of the telephone conversations is not being examined, only the network of interactions, there is no way of knowing if a particular contact of a known terrorist is a comrade in arms or just the local pizza outlet where they order a meal every Friday. Regardless, the fact that the mobile telephone facilitates direct contact between individuals and the fact that it can be used anywhere and anytime mean that those with both good and bad intentions can use it to further their aims. The electronic traces left by these interactions are of interest when trying to determine whether a particular person was at a murder site, to halt terrorists or to stop potential coup participants.

4 Mobile communication, rule transgression, quasi-legal and illegal activity

There are laws, there are rules and there are agreements. Laws are formalized rules (they have been established through some procedure) and they are binding – for example, in many places it is against the law for teens to drink alcohol. There are also sanctions associated with breaking laws. A rule is much less stringent in that it is only a general guideline. The weakest of these is an agreement – that is, a type of informal accord reached between individuals, based on mutual understanding.

The rise of the mobile phone has led to the establishment of rules regarding its use. As with other forms of technology – the TV, the PC, calculators, etc. – we have had to work out when and where it is acceptable to use it. For example, when it was first developed, there were extended discussions with regards to the use of the calculator in the classroom. These discussions had the tone of moral arguments.

We have seen the same discussions regarding use of the mobile phone in schools. School is a location where there is generally a rule against use of the mobile phone in the classroom. In that setting the focus is on teaching/learning, and the use of the mobile phone is often seen as being disruptive (Campbell, 2006). In addition, it threatens the notion of the individual scholar doing his or her own work, in that the mobile phone allows for various forms of cheating and deception (Ling, 2000). Needless to say, these rules are not always respected (Ling, 2009). In a 1999 survey of 1,000 teens in Norway, at a point when about 66 percent of teens owned a mobile phone, the material shows that just fewer than 20 percent reported having received a text message in school on the previous school day.[12]

Perhaps it is therefore not surprising that the mobile phone, and in particular texting, have been employed for "assisting" students with exams. Two of the most famous cases were at the University of Maryland, College Park, in the USA, and Hitotsubashi University in Japan. In the University of Maryland case, the inventive professor posted a bogus answer key on the internet after the exam had begun (Katz, 2005). Fellow students who were not taking the exam sent the information to their friends who were taking it. About a dozen students were caught in the sting. In the case of Hitotsubashi University, 26 students were failed when it was discovered that they had arranged to receive answers for an exam on their mobile phones.

4.1 Chicanery and bullying

Another somewhat sophomoric form of deviance is the use of the mobile phone for different types of chicanery. The mobile phone – and now the camera phone – can be used for various forms of traditional chicanery. With landline telephony, this often consists of obscene or inappropriate calls. There is a certain ambiguity regarding the use of landline telephony for these calls since several people within the household often share the same telephone. This ambiguity is eliminated with the mobile phone (Katz, 2005) and, indeed, it is an increasingly common problem. It has been reported, for example, that 11 percent of women in Germany received illicit calls via their mobile phones (Kury, Chouaf, Obergfell-Fuchs & Woessner, 2004).

The mobile phone – often in its guise as a camera / photo distribution device – is also used to take inappropriate photos and to distribute them. These can include surreptiously taken "gym shower" photos, or more simply photos that catch the subject at an inappropriate moment. Camera phones can then quickly transmit the photos and even post them on the internet for a potentially global audience (Brandtzæg, 2005; Tikkanen & Junge, 2004).

Chicanery is disturbing to the victim if it happens once. If an individual is repeatedly the victim of bullying, it becomes a serious situation (Ling, 2007a; Tikkanen & Junge, 2004).

Mobile communication provides people with a medium through which the bullying can easily be communicated. Again, since the mobile phone is often seen as a part of our personal kit, it also provides others with a direct link through which they can send offensive messages remotely. The advantages – as seen from the perspective of the person doing the bullying – are that he/she is more anonymous and there is virtually no possibility of supervision of interactions as there is in classroom or playground situations (Patchin & Hinduja, 2006). Indeed, the victim of so-called "cyber bullying" cannot seek refuge at home or in other locations since they are ostensibly available to be victimized whenever and wherever. In an experiment designed to test the impact of ostracism communicated via text messages, Smith and Williams found that even a mild form of exclusion can result in a reduced sense of belonging and lower self-esteem (2004).

Patchin and Hinduja reported that about 2 percent of their sample had been bullied. When looking at the previous 30-day period, the subset of those who had been bullied via the mobile phone reported a median of 3.3 events – that is, an episode about once every 10 days. One poor soul reported 23 events in the 30-day period. By comparison, internet-based bullying was more common in the material reported by Patchin and Hinduja. Almost 30 percent of their informants had experienced some form of online bullying.[13]

There are gendered differences with adolescent males reporting more bullying than females (Kowalski & Linber, 2007). Raskauskas and Stoltz show that the people who bully and those who are victimized in cyber bullying are generally the same as the people who take these roles in traditional co-located forms of bullying (2007). In addition, victims often did not know the identity of the person initiating the episode (Li, 2007).

4.2 Status offenses among teens

Another form of criminality are the so-called "status offenses." There are some forms of deviance that are illegal regardless of the age, gender or status of the individual (murder, theft, etc.). There are, however, some forms of deviance that are only considered illegal for persons of a certain status. Generally, these are things like underage drinking, sexual interaction before the age of consent, etc. These transgressions often become a part of teen boundary testing as they move from the relatively stable haven of their parents' home into the different currents of adult life.

Analysis has shown that teens who are heavy users of mobile communication often are overrepresented among those people who also engage in a variety of illegal activities. The results are perhaps most striking when considering underage sexual activity. Analysis of material in Norway has shown that the more a teen of a particular age uses their mobile phone, the stronger the possibility that they will have had their sexual début (Ling, 2005a; Pedersen & Samuelsen, 2003).[14] There are several interesting issues here. The careful reader will note that we have not posited a causal relationship between phone use and sexual activity. The actual form of the link between the two is not understood. There are two possible issues here, however. The first is that those teens who are most gregarious and social both like to use the mobile phone a lot and are also the most adventurous sexually. Another issue here is that the mobile phone is also a tool of coordination between individuals. Thus, it is a perfect device for arranging a discreet tête-à-tête.

4.3 Illegal activity

There are also situations where the mobile telephone has facilitated illegal activity. Just as mobile communication assists the functioning of licit activities, it also helps those who are bent on carrying out illicit ones. The same analysis that showed the relationship between mobile phone use and sexual activity among teens also showed a much darker relationship between heavy telephone users and various types of serious criminality (Ling, 2005a). The data showed that heavy mobile phone use covaried with a variety of illicit activities such as breaking and entering, stealing, vandalizing and fighting with weapons (usually knives).

Women who are trying to escape abusive relationships find that the use of mobile phones and PC-mediated interaction make it more difficult to break out of the relationships. Their abusers – often former husbands or boyfriends – can harass and threaten the women in a type of psychological abuse (McKenzie-McLean, 2007).

The mobile phone is sometimes used for various forms of fraud and electronic sabotage (Leavitt, 2005). The adoption and use of mobile communication has altered the sales channels for drugs such as heroin in the New York City region (Furst et al., 2004), it has facilitated mobilization for gang fights in Norway and Australia (Carlsson & Decker, 2005; White, 2006) and has aided in the organization of prostitution in Vietnam and Bangkok (Doussantousse & Tooke, 2002; Plant, 2002). Varbanov (2002, 123–33) describes the use of mobile phones among criminals in Bulgaria immediately after the fall of the Soviet Union. In this case, mobile phones were used to organize various types of criminality including extortion, theft and gang-style killings.

Earlier, we described how mobiles help protesters arrange demonstrations; they can also aid in more direct, harmful and violent forms of opposition. The most obvious example is their use to trigger bombs (Cumming-Bruce, 2005). As noted in chapter 4, mobile phones were also used in the organization of the September 11 attacks (Dutton, 2003; Katz & Rice, 2003) and they have been used to help survivors in the wake of bombings (Cohen & Lemish, 2005). Yet mobiles clearly cut both ways, helping and hindering terrorists and terrorist-hunters alike, depending on the situation. While at least one government minister in Germany wants to ban terror suspects from using mobiles at all,[15] in Afghanistan the Taliban has blown up mobile towers (masts) which they claimed were being used to track their movements.[16]

5 Mobiles as symbols or vehicles of globalization

Drawing back from the abyss of illegal activities, another debate associated with the mobile telephone focuses on its role in globalization. Mobiles are a particularly global technology. A user in Kenya might own a handset made by an alliance of a Swedish and a Japanese company. This handset was assembled in China, from materials that come from around the globe, uses a network protocol developed by an international consortium, and runs on a network funded by a joint venture between a Kenyan firm and an international telecommunications carrier based in the UK. However, this international flavor has another dimension – its significance varies from place to place: specifically, it interacts with the powerful imagery of local and international advertising, and with issues of national identity.

These international crosscurrents are so prevalent, the advertising messages so powerful, and the take-up of the technologies has been so rapid that publics around the world are racing to make sense of what using mobiles means to them and to their nations, vis-à-vis the rest of the world. As we mentioned earlier, mobiles are marketed as stylish, desirable consumer objects. Thus, to the established dynamics of the modern (vs. the traditional) and youth (vs. age), we add a third debate; the local vs. the global, us vs. them. The meaning of the mobile is caught up with the broader discussions about modernization, urbanization and, particularly, globalization. A brief spin around the globe reveals the following examples, which range from the optimistic to the dystopian.

We can start with a re-acknowledgment of the importance of mobile phones to hundreds of millions of people living away from the countries of their birth – the world's economic diasporas (GSM Association, 2007). Amihan, the Filipina maid working in Singapore, is an actor in a global system that has grown up alongside inequalities in standards of living between countries. She can earn more doing the same work in Singapore than she ever could in Manila, so she stays overseas, speaking to her family mostly, these days, by mobile phone. Serigne Tall describes how the mobile's introduction into Senegalese villages has accelerated a kind of informal globalization, by allowing closer social and financial relationships between residents of some villages and their émigré counterparts in cities in Italy than between those villagers and people (and officials) in the major cities of Senegal. People are making the connections they want to make, and their use of technologies is outpacing use by the state. Tall notes that this is evidence of a "destabilizing effect on the classical order" (2004, 47).

Yet mobiles need not actually connect people in conversations across borders to play a role in globalization. A longtime observer of mobiles in the Philippines, Raul Pertierra, talks about the symbolic role of the mobile by suggesting that it has replaced the jeepney (refurbished Second World War jeeps that served as ubiquitous means of public transport) "as the expression of what is quintessentially contemporary Filipino. Both icons deal with mobility and connectivity, but while the jeepney traverses physical distance the cell phone supersedes it" (Pertierra, 2006, 9). Similarly, Varbanov talks about the spread of mobiles in Bulgaria, against the background of the collapse of communism and Bulgarian society's democratic and economic transformation. To him, the mobile's "status as a national icon seems assured, and its symbolic and phenomenological prominence speaks volumes for the nature of culture and hierarchy in Bulgarian Society at the start of a new millennium" (Varbanov, 2002, 136).

Writing about India, Kavoori and Chadha (2006, 234) suggest that, thanks partially to glossy advertising rife with appeals to particular western sex roles, the mobile "frames and constructs Indian consumers along a semantic/narrative story line that centers individuality, references corporate-sponsored identity formations, and is reiterative of very specific modalities of identity formation in the Indian context in the wake of market liberalization" (see also Mazzarella, 2003). Zhao (2004) discusses *Shau Ji*, a Chinese film that views the mobile in a dystopian light and as a tool likely to destroy individuals and families. Here the concern is not so much western or foreign values as urban values – the protagonist is a yuppie who finds his life and relationships undermined by the mobile.

This is not to say that backlashes are particularly common. These observations about negative or problematic linkages between mobiles and globalization remain generally in the domain of cultural critics. Very few institutions, save perhaps the governments of Cuba and North Korea, and the Taliban,[17] have both the power and the desire to try to keep mobiles out of the hands of ordinary people (and even the North Koreans and Cubans seem to be cracking).[18] Yet perhaps these debates play out around the intergenerational dinner table, or in the editorial pages of local newspapers. In practice, the debates about whether mobiles undermine or represent national identities are intertwined with the other debates about power relationships, intergenerational misunderstandings and unwanted disruption.

The backlash can also be seen in discussions regarding the effects of electromagnetic radiation and the issue of cyber junk. As Adam Burgess documents in *Cellular Phones, Public Fears, and a Culture of Precaution*, there has been a great deal of attention paid by the press, the public, policymakers and the research community to the health risks from radiation, electricity and other disturbances from handsets and cell masts (Burgess, 2004; see also Goggin, 2006). This debate is far from conclusive, but is a part of the broader discussion of mobile communication.

More recently, another group has begun to examine the mobile phone manufacturing and consumption lifecycle. Billions of handsets have come to us with costs which might not be reflected in the retail price of the handsets themselves. These externalities range from the social and political impacts of the rough conditions surrounding the Coltan (columbite-tantalite) mines in central Africa (Montague, 2002) – coltan is a metallic ore used in the production of most handsets – to the environmental problems of recycling and disposing of hundreds of millions of handsets per year, each loaded with lead, PVC, mercury, cadmium, barium and other nasty stuff (Carroll, 2008; Goodman, 2004).

Katz and Aakhus' theory of apparatgeist (Katz & Aakhus, 2002) suggests that any concerns that the guardians of the old, the collective and the particularistic have about mobiles may be appropriate – that the problems are not only with the way mobiles are marketed but also how they are used. If, as apparatgeist suggests, mobiles do have an essence or "socio-logic" which enables a universal shift towards "perpetual contact" among their users, then there are certainly institutions and actors who will be threatened by the device, and others who feel it is their responsibility to protect established orders, societies and norms from incursion by mobile telecommunication. The "classical order," as Tall puts it, may sense a threat.

6 Conclusion

In the previous chapters, we have seen how people around the world are using mobile telephones in daily life. In this chapter, we have focused on some of the unintended – and perhaps unwanted – consequences of mobile use. The topics of the debates associated with the device include things both mildly irritating, like the disruption of co-located interaction, and deadly serious, like the mobile's use as a way to detonate bombs.

In many ways, the mobile telephone confronts the existing order. Use of the device plays into the way that social protests are organized and the way that citizens band together for various purposes. The mobile phone changes the equation in relation to news-gathering and it also affects the emancipation process and the attendant negotiations and demands that take place between teens and their parents. The mobile phone also provides underhanded authorities with new ways of gathering information and performing surveillance. The mobile phone provides new avenues for the commission of quasi-legal and patently illegal activities. Everything from chicanery and bullying, through the organization of theft, murder and extortion, to the commission of terrorist acts can be supported with the use of mobile communication.

Finally, the mobile phone is a particularly strong example of globalization. The production of the mobile phone systems as well as the devices draws on many parts of the world. The companies that market the handsets and operate the networks are, as often as not, multi-national concerns. Mobile phone handsets often have a life that starts in a comfortably well-developed economy, but, when the style turns to the next phase of handsets, the formerly *au-courant* handset is many times sent off to a developing country, where it is recycled and sold to an appreciative second user. In this capacity, it may facilitate communication between a family which has a son, or a father or a mother, working overseas, who uses the device to send home money in the form of remittances. When the handset has ended its useful life, it becomes cyber junk and may be broken up for its component parts and materials. These issues, just like the widespread adoption of the mobile phone, are all a part of the bigger picture.

CHAPTER SIX

Conclusion: individual addressability, interlacing and the spillover of the control revolution

1 Introduction

Earlier in the book, we introduced ten characters, and illustrated the ways in which the mobile makes a difference in their lives. It provides basic connectivity for those who have not experienced it before, and it facilitates a world of instrumental and expressive interactions to all who use it. But a different perspective on the experience of those same people would suggest that, at an abstract level, not much has actually changed. We were coordinating and getting help, flirting and reminiscing, buying and selling, and generally engaging in the exigencies of daily life long before mobile communication arrived on the scene. The mobile phone has simply facilitated these processes.

So which is it? Does mobile communication signal a revolutionary new era, or is it just a nifty tool with which we do the same old stuff? As with many things, we want a clear picture, but the reality is more complex. The addition of over 3 billion mobile subscriptions in a decade is a monumental shift, experienced particularly in the developing world. Access to basic telephony has never been easier, less expensive or closer to universal. By contrast, the transition is perhaps not so dramatic for persons in the developed world, for whom mediated interaction has been on the scene, readily accessible in most homes and workplaces, for some time. Nonetheless, mobile communication presents a shift, albeit more evolutionary than revolutionary. The mobile phone is a complex and powerful device. It is both a tool and a symbol. It is a gizmo, a harbinger, a conduit to loved ones, an actor with its own sense of agency, a savior and a curse, a tool and an expression. A steady growth in functionality has modified and extended the device beyond its core capacity to move our voices over distances. Through images, video, text, GPS and data, it is becoming another path to the variety of content, services and communities on the internet. It is a force for economic development; it can extend education beyond the classroom and provide access to information regardless of location. It is being used by teens to distance themselves from their parents, and as a way for parents to keep track of their offspring. The mobile telephone is simultaneously a consequence of social processes and a device that has social consequences. It is so varied in its application and use as to render a single, linear, global, unified explanation of its impact impossible (and perhaps undesirable). In sum, there are many different applications of the technology, each co-evolving alongside pre-existing forms of human interaction.

Mobile communication, however, has essential attributes, which distinguish it in form and impact from traditional landline telecommunication. First, as *mobile* handsets become *ubiquitous*, the mobile phone provides for individual addressability – we place calls to people not places, and expect to reach those people when we call. Second, the same ubiquity means that, as we carry our devices everywhere, we can interlace our communication activities more finely into the fabric of our everyday interaction in a way that was not previously possible. By interlacing in this way, we attempt to control the flux of activities buffeting our lives.

Further, widespread access, individual addressability and the ability to interlace interactions means that we have introduced a type of *mobile logic* into our interactions. Around the world, there is an emerging norm of connectedness, an assumption that all of us are available via a mobile phone. This is not merely an expectation that we can bother Bill or Sally in an emergency, but rather something deeper and habitual. This chapter will argue that, thanks to an emerging mobile logic, our everyday actions are determined, at least in part, by our expectation that others we may interact with are always available via the mobile phone; thus, planning is different, travel is different, togetherness is different, and privacy is different.

2 Individual addressability and interlacing

In 1972, the evening news in Norway broadcast a story about an early "mobile telephone" system. The voice-over described the development of the system, the plans of the state-run telecom operator and the anticipated number of subscribers (there were some 40 base stations for the whole country and 2,100 people were subscribers). The accompanying visual images opened with a fellow in a suit and tie speaking into a walkie-talkie-like microphone while holding a device that is roughly the size of a lunch-pail. This image faded and was followed by that of another man (also in suit and tie) shown running across a lawn while speaking into the same mobile telephone. The viewer was led to believe that he was in the process of reporting some emergency and was in great need both of getting out the message and, at the same time, of fleeing (or perhaps he is rushing to give assistance, the plot line isn't well developed on this point). The text and the subtext of the story are clear. At the overt level, the viewer was being informed of new technical developments. The visual presentation in the story is, however, prescient. It first introduced the viewer to the device and then introduced the notion that it will allow us to call (and be called) regardless of where we are. This story also developed the notion that the mobile telephone would allow us to intermingle different activities. Seen with these eyes, the news story serves to illustrate two of the basic functions of mobile communication noted above, namely individual addressability and the interlacing of activities.

2.1 Individual addressability

When we use a mobile, we call or send a text to a person, not a place. Each of us becomes individually addressable, and that makes all the difference. When using the traditional landline-based system, we "come to the phone." Indeed, a polite way of refusing a phone call was "She cannot come to the phone right now. Can I take a message so she can call you back later?" The statement assumes that the telephone terminal was fixed and we moved in relation to it. With landline telephony we called fixed locations in the hope that a particular person would be nearby, and that they would not be indisposed at the moment. Clearly, these are bygone niceties. Having a mobile phone – along with caller ID and voice mail and texting functionality – means that we can call or text directly to the individual. If we are "indisposed" at the moment, we can still see who is calling and make a snap decision as to whether we should end our indisposition in order to take the call or simply send it to our voice mail account. The problem is even easier with texting, which is by its nature more asynchronous. With these technical fixes we do not have the need for artful excuses when we do not want to take a call – though we may need them later when confronted by the spurned interlocutor.

The mobile phone puts each of us into play. We are remotely accessible to others. Parents can call teens, patrons can call local restaurant owners, businesspeople can call colleagues, etc. If we put this development into the context of mediated interaction, mobile communication represents something new in several different ways. It is interpersonal and it is interactive. As suggested by Ito, Okabe and Matsuda (2005), it is also a personal medium. The mobile phone has, in many cases, become an object we have on our person and so the communications come directly to us.[1] The ability to talk or to send a message to people, regardless of where they are, is a fundamental shift. Thinking historically, with the telegraph – as well as different pneumatic message delivery systems – messages were delivered between central offices and then taken to an address by hand. The landline telephone expanded the reach of the network into the home and office. Most often, a common phone was shared by many people (e.g. in some offices, dormitories, etc.). Indeed, the phone booth was a type of socialized telephone that could be used by anyone who needed to make a call when they were away from other possible phone terminals. The mobile phone has changed this in that those people who have a mobile phone are personally accessible whenever and wherever they happen to be.[2] There is also an increasing tendency to replace landline telephones with mobile phones. In the USA, for example, the proportion of people who have only a mobile phone, and not a landline connection, has gone from about 4.4 percent of the adult population to 12 percent, between 2004 and the end of 2007 (Blumberg & Luke, 2007). Thus, the tendency is clearly that we are moving towards the personalization of telecommunication.

2.2 Interlacing of activities

The potential for individual addressability allows for the interlacing of activities. With mobile communication there is an intertwining of mediated and co-present activities (Ling, 2008; Taylor, 2005). We call a friend to chat while waiting for a bus. Students like Stan can send and receive text messages during a particularly boring lecture. Office workers like Mette can check on their children in the time between meetings. That is, we engage in mobile communication in the various folds of life when we are not otherwise engaged. The act of communication with remote persons (or in some cases data sources) has become more integrated into our everyday lives. These interactions are taking place as needed and they are taking place between specific individuals. Licoppe discusses the way that mobile communication, and in particular texting, allows for what he calls "connected presence" (Licoppe, 2004). The idea is that we carry the interaction with our friends and family through the day. At one turn we are able to send them a message, and in the next turn we meet them for a cup of coffee. The next interactions might be a short series of calls to lay further plans, and then there might be a couple of text messages to confirm things or perhaps to share a small joke or a bit of gossip. There is no need to set aside time for longer involved discussions; rather, we interact with them in fits and starts through the day. There is a more-or-less continual stream of interactions that we fit into the open moments of our daily routines.

Interlacing of activities means that we can vicariously participate in situations or we can organize one social interaction while we are still (physically) in a separate one. While in class, the student can send messages to classmates regarding their upcoming meeting in the library to work on a class project. The lover can send messages – or even missed calls – to the object of his/her love while at work or waiting for a bus. In the middle of a meeting, we can receive a call from our child's school telling us that our child is sick and needs to be picked up. The different activities, impulses and situations are being worked out simultaneously. One situation is being attended to while we are physically in another. It is not just making concrete arrangements but also sending messages and making calls that are intended to express our feelings and to warm the soul of our interlocutor. The mobile phone allows us to wedge quick messages into time we spend waiting for other activities to commence.[3]

3 The control revolution in the social sphere

Individual addressability and the interlacing of activities affect the way we coordinate our lives. When seen as a mass phenomenon, they also suggest that there is a changing logic in the organization of interaction.

There are also those who suggest that the rise of mobile communication is perhaps both a cause and a result of more haste in society (Arnold, 2003; Gripsrud, in progress; Virilio, 1995). The idea here is that society is characterized by a more rapid tempo. James Beniger (Beniger, 1986) examined some of the same issues, portraying progressively increasing speed in the form of industrial production and commercial interactions. This result, he argues, has been a crisis of control. Using the most dramatic of Beniger's examples, the speed of the train system demanded that there be a control system in order to prevent the trains from occupying the same tracks and crashing into each other. It was for situations such as this that ICTs, such as the telegraph, were designed.

In contrasting the industrial revolution with what he calls "the control revolution," Beniger writes: "By far the greatest effect of industrialization . . . was to speed up a society's entire material processing system, thereby precipitating what I call a crisis of control, a period in which innovations in information-processing and communication technologies lagged behind those of energy and its application to the manufacturing and transportation" (Beniger, 1986, 427). As a response to this, organizational, mass media and telecommunications systems were developed. Writing before the rise of the internet and the mobile phone, Beniger cites the impact of both telegraphy and traditional landline telephony as examples of telecommunications systems that were a part of the control revolution. His focus is consistently on the interaction between production processes and the systems employed to control them.

Telecommunication unquestionably helps to control the way that industrial and commercial processes proceed. Mobile communication follows in this spirit – and indeed it was spawned from the industrial process in association with the electronics industry – in that it has allowed new forms of coordination and it has made the production of various goods and services more efficient (Julsrud & Bakke, 2008). With all deference to Beniger, however, that is only half the story. Beniger never makes the transition into the impact of ICTs on the social sphere in his analysis of the control revolution. As we have written here, mobile telephony has lowered the threshold for social interaction. Just as with the internet, the technology is facilitating a range of interactions that have an exclusively social profile, including an extension of control or coordination.

If we accept the assertion of Beniger that communication technologies such as the telegraph, the telephone, the internet and the mobile phone arose in the service of controlling production processes, we can nonetheless see that they have been widely applied to non-commercial purposes. Historically, the telegraph (Standage, 1998; Stevenson & Bartunek, 1996) and the telephone (de Sola Pool, 1977; Fischer, 1992) have also each had a dual identity as both a system in the service of commercial interests and, at the same time, a conduit for social interaction.[4]

The speed with which society is moving may well be a part of the picture. Thinking about the commercial side of life, there is a seeming haste to many of the commercial imperatives that stress, for example, "shortening the time to market" and other similar slogans of the business world. This said, we are not necessarily lost in the rush. Rather, the notions of individual addressability and the interlacing of activities can also mean that we are able to hold open several streams of interaction simultaneously. The small entrepreneur can be using the mobile phone for placing orders to a remote supplier and, in the next turn, get an order from a local customer followed by a missed call from a friend. The business executive can be organizing a meeting via the mobile phone and checking on the status of a project in the same turn of events, and then get a text message from his daughter reporting on her success in a math exam at school. The student can be attending a lecture while covertly arranging to meet with friends via SMS, the harried mother can use the mobile phone to organize driving to after-school activities and the father *qua* soccer coach can send a text message to his team to remind them of when and where they will practice. However, access to a mobile communications device means that we can carry out these tasks as needed in the folds of life. We can contact others when we have a short gap in the co-present flow of events – or when the lecturer is boring. That is, we can have a series of short interactions to arrange something or we can exchange a series of ritualized comments (endearments, a joke, etc.) and then the phone goes back in our pocket or purse.[5] While the opposite point can be argued – and indeed is effectively argued by, for example,

Chesley (2005) – mobile communication can also help us to hold the increasing tempo of society at bay.

3.1 Control spillover

Mobile phones have been adopted around the world for commercial as well as explicitly non-commercial purposes. The devices are still being used by executives and the heads of industry. But, perhaps more profoundly, they are being used by teens in the suburbs, farmers in Africa, mothers in Germany, just as they have been adopted by financers in Argentina and fishers in India. It is possible to assert that we are in the process of a "control spillover," in that control processes that were engendered in an industrial context have been democratized and applied to interpersonal interactions. The mobile phone (in its role as a personalized networked device) is the most obvious of these control systems. It is a way to interact (and perhaps control) others ("When are you coming home for dinner?"). It is also a way to do nurturing and to carry out ritual interactions. Thus, it is the application of a control technology to the exigencies of our personal lives.

As we have noted, there is widespread adoption of mobile communication and it is being used in the processes of everyday life. We have also suggested that this does not necessarily mean that we are doing wildly new things with the mobile phones. Rather, we are doing many of the same things as before, albeit more efficiently or perhaps more quickly. We are applying the individual addressability of the device and the ability to interlace communication to the issue of controlling interactions. We are basically using the potential of mobile communication to coordinate with others and to communicate about everyday affairs. There are new forms of interaction such as texting, but, even here, the content of the messaging is often about the affairs of everyday life.[6] Messages such as "Where are you?" and "Can you buy some milk on your way home?" are common (Ling, 2005b; Ling & Baron, 2007). Indeed, it is possible to say that the mobile phone is a technology of the intimate sphere. The main focus of mobile communication has until now been interpersonal interaction with a limited number of people. The great majority of calls and text messages are actually within the sphere of family and friends (Ishii, 2006; Kim et al., 2006; Ling, Yttri, Anderson & DeDuchia, 2003; Miyata, 2006; Reid & Reid, 2004; Smoreda & Thomas, 2001; Wei & Lo, 2006).

We are engaging in what Licoppe calls "connected presence" (2004). With a bow to Townsend (2000), we might also call it "real-time networking." The evidence up to this point indicates that mobile communication facilitates use of the links within the primary sphere (family and friends) (Ling, 2008). While we might have the image of a mobile user as a high-flying executive who is seemingly in touch with the world, the reality is that the majority of most people's calls go to only a small number of other individuals.[7]

3.2 The mobile logic of the real-time social sphere

It is possible to suggest that, just like the automobile before it, mobile communication has become a taken-for-granted aspect of everyday life. Upon its introduction, the car was an oddity for only the rich. There were no adequate roads, there were no gas stations, and nor were there mechanics who could repair the vehicles (Fischer & Carroll, 1988). By now we realize that – for both good and bad – everyday life in much of the developed world has been restructured by the automobile. Jobs, education, entertainment and the simple mechanics of life often require automotive transport, and even those people who do not have a car must organize their lives around a system that is organized for those people who do. The automobile and automobile-based transport is a taken-for-granted part of our lives. In the meantime, millions of new middle-class families in India, China and throughout the developing world are purchasing their first low-cost automobiles! Just as roads, suburbs and shopping malls structure modern societies, so too does mobile communication. The arrival of a mobile logic suggests that we arrange our daily affairs with the assumption that one and all are available via mobile communication.

We see this perhaps most clearly when confronted with a breach in everyday routines. These can be minor events, such as having to rearrange meetings when we find ourselves sitting in a traffic jam. Or they can be personal emergencies where there is the need to quickly rearrange our lives. Not long ago, the first author of this volume was involved in a minor emergency that illustrates the way that the web of social interaction is facilitated when using mobile telephony; it also shows how the mobile phone has become a taken-for-granted part of daily life:

> A woman fell on a stairway and hurt her leg when she was rushing to get her groceries into her apartment while her two-year-old son was asleep in the back seat of the car. As is the case in situations such as this, there were a whole series of issues that needed to be dealt with. Aside from a banged-up leg and the resulting shock, the woman was not otherwise hurt. To confirm this, however, she needed to go to the emergency room. In addition, her son needed to be cared for. Thus, there were a whole series of communications to be made.
>
> When lying on the stairs, before other assistance had even recognized that there was a minor emergency afoot, the woman had incidentally received a call from a friend who had rung for a chat. After being alerted to the situation, this friend was on her way to the apartment but was still a half an hour away. After this call I chanced by and was drawn into the situation. I helped her to a more comfortable position, she was able to call another family member to come and help with taking care of her son. This family member was en route to another location at the time but it was arranged that he could come and get the child, but it would take approximately an hour. A short-term babysitter was found – my daughter. In addition, another friend was alerted and he was able to meet the woman at the emergency room after he had retrieved his own child from day-care. Finally, a call was made to my wife in order to postpone my picking her up from a shopping trip.

The importance and sequence of the mix of calls to different corners of the city in this situation brings us back to an
observation by James Katz: "You're a problem for other people
if you don't have a mobile phone" (Weiner, 2007). If one of
the individuals involved in the incident of the woman falling
on the stairs had not had their mobile phone, it would have
resulted in a less efficient resolution of the situation. Clearly,
the woman would have eventually been taken care of without
the mobile phone. However, the device was a central link in
the coordination of the various efforts.

The interesting thing in all of this is the efficiency of the
interaction. People were alerted and activities were put on
"hold" through a series of calls that were routed directly to
them. Had there been only a landline system, we would have
had to rely on calling others who then would have to relay messages to others while awaiting the arrival of people at a variety
of destinations. Instead, those people who were expecting to
be picked up could be notified that an urgent situation had
arisen and that it was better to find an alternative. Underlying
all of this was the assumption that each relevant person had a
mobile phone and was accessible via that form of mediation.
This assumption has become a part of the logic of a real-time
form of coordination. Relying on this logic, the connections
were made, the issues were coordinated in what approached
real time, the woman was driven to the emergency room, her
son was taken care of after a short interlude with a babysitter,
and the groceries were put away by a good-willed friend.

Not all our interactions are as dramatic as having to deal
with an emergency situation. Both of the authors have found
other indications of the arrival of a mobile logic in our interviews with teenagers. For example, when asked about not
having a phone, teens in Norway said the problem is more
for the people trying to call you to sort out. If you want to
place a call, it is easy to borrow a phone. Someone always has
one. You can get word out. However, if someone needs to call
you and you do not have a phone, then the caller has no way

of completing his or her mission. We found a behavior to make up for this problem, among middle-class Indian teens – among whom ownership rates are climbing but not nearly as high as in Norway – which we called "approxi-calling." Approxi-calling is a strategy to reach someone who you know does not have a mobile of their own. The teens we spoke to tried to locate their phone-less friends probabilistically – based on the time of day and a knowledge of their friends' habits and associates, they call the mobile they think is likely to be nearby – usually that of a mutual friend (Steenson & Donner, 2009).

These glimpses of situations and strategies, whether mundane or urgent, point to the different social metabolism associated with the coordination of activities via mobile communication. Anthony Townsend introduced the notion of the real-time city where mobile communication facilitated, for example, the efficient dispatching of taxis through informal interactions among the drivers themselves (Townsend, 2000). The idea is that mobile communication has also become a system through which we work to control the tempo of our lives. Just as the city has grown with and been shaped by the automobile (Crawford, 1994; Dyckman, 1973; Grahm & Marvin, 1996; Hall, 1996; O'Connor & Maher, 1982; Thorns, 1972), the mobile telephone is in the process of reforming the way we coordinate interactions. It ties groups of individuals together in a way that has not been possible before. In addition, it is in the process of creating a logic of interaction based on personal accessibility.

3.3 Are things going faster (or too fast)?

There is a tension and a kind of Janus face to the technology (Arnold, 2003). On the one hand, the mobile phone is the child of the control revolution. Following the line of Beniger, it is a device that is being used to deal with the increased complexities associated with industrialization. It is a way to facilitate coordination since, to use Townsend's phrase, it allows for real-time interaction (2000). It is a way for us to somehow hold at bay the stresses associated with ever more voracious jobs and social exigencies. At the same time, it is a tool of those same impulses. It is a way that our bosses can get in touch with us at untoward hours. It is a disturbance when we want to simply relax and not think about all the incomplete or pending assignments that await us. Hochschild describes this as the invasion of private time with work (2003) while Hylland Eriksen examines the general stressing of society (2001). There are both pressures and the release from pressure in the device. To say it is one or the other is to not recognize the whole picture. Just as in industrial production, many have suggested that there is also a greater speed of social interaction.

As noted above, Chesley suggests that the mobile phone facilitates "work to family spillover" that plays out, particularly in the case of women, in terms of increased stress and lower family satisfaction (Chesley, 2005). Her results point to the Faustian dilemma. In the hope of gaining a type of control, we are adopting the tools that allow further speed-ups.

At the same time, the dispatching of small tasks and dealing with different exigencies may also open up time. We might be able to deal with the small incidentals via the phone and thus clear up spaces that would otherwise have been cluttered with "the small stuff." A text message from our husband/wife saying that we need milk can mean that we are able to save time and effort in what would otherwise be a second trip to the store. In addition, were we not to have a mobile telephone available, we would be out of the loop vis-à-vis our social group, be it in the case of an informal gathering or of some emergency.

Almost without realizing it, we have moved into real-time logic, where there is the expectation that everybody is reachable. On the one hand, there is a sense that the tempo is increasing. However, there is also the notion of Beniger that the technology provides at least the hope of control.

4 Four billion mobile phone users

The first decade of the twenty-first century is yielding one of the greatest explosions in connectivity ever experienced. As we discussed in chapter 2, the end of 2008 saw the activation of the 4-billionth mobile phone *subscription*. After accounting for unused subscriptions, and for those among us who carry more than one SIM card or mobile phone, it is still the case that sometime very soon, perhaps by 2010, someone will become the 4-billionth mobile phone *user*. Of course, this 4-billionth user might live anywhere, in Los Angeles or Munich or in a suburb of Osaka, but it is more likely that he or she will live somewhere in the developing world. From the tiniest hamlets in the smallest countries to the crowded streets of the mega-cities of mega-nations of the developing world, the people of our planet are still lining up to buy mobile phones. These billions of new telephones on the planet have implications for individuals, for societies and for the human community as a whole.

At the individual level, the people and families who purchased these handsets will have access to all the advantages and disadvantages of mediated communication; just like those lucky folks who have a landline, the new mobile-only users can call for an ambulance if someone is sick, check on the price of potatoes, call their mother, call a radio talk-show or just talk about the weather. Many of them will struggle to set up voice mail, some will screen their calls or tussle with their kids about how much to spend on telephone calls. Bad guys will plan bad things using their new telephones; good guys will plan good things, as well. The mobile is, first and foremost, another telephone handset, and another way to access the global telephone network.

However, as the stories of Liang and Annette, and the other vignettes from chapter 3, illustrate, the new "mobile-only" users are going to be wrestling with all the same debates around mobile communication as their compatriots who are adding a second or third mobile to the landlines in their homes. When

Liang gets a text message from his friends, they are expecting to reach him directly, not the hostel where he lives; Liang had to wrestle with individual addressability from the first day he had access to a telephone of his own. When Annette's customers call her restaurant to order lunch, they might already be on their way, walking or riding a two-wheeler towards her shop. Her customers, who are likely to be mobile-phone-only users, have adapted to an "interlaced" day almost as soon as they purchased their handset.

At the societal level, the sudden increase in affordability and accessibility of mediated communication brings more societies closer to the promise of "universal access". The Maitland Commission Report in 1984 revealed that two-thirds of the world's population had no access to telephone services (see also Hills, 1990; Maitland, 1984, 13). More recently, Zainudeen et al. (2007) studied five countries in Asia – Pakistan, India, Sri Lanka, Philippines and Thailand – and found that, at least in the Asian context, something over 90 percent of their 8,689 respondents had used a phone in the previous three months. Using one of the metrics often associated with the Maitland Report, Zainudeen et al. found only 5–10 percent of people in the different countries lived more than 60 minutes' walk away from the nearest phone, and in all cases 80 percent had access within a 5-minute walk. And these booms are not just in the developing world; in an analysis of Norwegian teens, for example, in 1997, only a very small number of teens reported owning a mobile phone. For example, only 2.5 percent of 13-year-olds, and 11.5 percent of 15-year-olds, said that they had their own mobile phone at that point. Two years later, in 1999, the number of 15-year-olds reporting that they had a mobile phone had risen to 68.5 percent, and, by 2005, it was literally not possible to find 15-year-olds without a mobile phone.[8] Thus, within a decade, access to the device had gone from being a rarity reserved for only a small number of persons to being an omnipresent part of everyday life. The stark gap that the Maitland Commission described is easing. By now, 2008, we might comfortably assume that half the world has made a telephone call; indeed, it is possible that, by the end of the first decade of the twenty-first century, half the world, or at least half its families, will own a phone.

Neither universal ownership nor even universal access will be achieved by the end of this decade. In terms of access, there will still be some remote villages, particularly in areas of low population density, where it will be economically unviable to maintain mobile phone towers. It is possible that some combination of regulatory interventions and donor or community support will find ways to provide coverage to these remote areas, but the problem itself is fragmented. There are still sparsely populated (but not unpopulated) pockets of the United States[9] and Australia[10] with people living in them and no cell phone coverage – how could we expect to achieve 100 percent coverage in similarly sparse or remote parts of varied landscapes like the deserts of Chad, the steppes of Mongolia or the rainforests of Brazil? But, on balance, the movement towards universal access is remarkably strong: by 2005, nearly 80 percent of the world's population already had access to a mobile phone signal outside their homes (World Bank Global ICT Department, 2005). Thus, the mobile logic of the real-time social sphere is spreading. Just as with the automobile, there will likely not be universal access, but the technology is undeniably moving in and establishing itself. It is becoming domesticated.

In terms of ownership, mobile telephones may continue to be outside the realm of affordability for those in the direst of circumstances. For the nearly 1 billion individuals living on a dollar a day or less (World Bank, 2007), the outlay for a $20 handset, or even a couple of dollars a month for access and airtime charges for a SIM card to put in a shared phone, may be too much, given all the other pressing requirements for survival (Hammond, 2007). Despite second-hand phones, shared phones, shared SIM cards and a host of cost-saving strategies (Zainudeen et al., 2007), the ongoing cost of maintaining access to one's own may not always be worth it. But as telephones get less expensive and more plentiful, the proportion of people who cannot imagine owning a phone of their own may begin to approach zero.

Thus, overall, the march continues towards more telephones, in the hands of more people, than ever before. These developments have implications for the world as a whole; there will still be countless ways in which the lives of the children of Mika (in Japan) and Annette (in Rwanda) will differ, but using a telephone will probably not be one of them. Does this make it a "smaller world?" In one sense, clearly yes, as more and more families which comprise the world's diasporas will use mobiles as Amihan and her daughter do, to stay in touch across great distances. However, it is not the case that Mika's kids and Annette's kids will be any more likely to speak than Mika and Annette were, prior to their mobile purchases. The world will still be big, and, as we described above, most calls will be local, or to close associates, or both. Nevertheless, thanks to globalization, Mika's and Annette's children are linked economically, culturally and environmentally, more than ever before. Economically, the rapidly increasing penetrations of mediated communication in the developing world will play a part in improving the productivity of businesses, governments and institutions (Waverman et al., 2005), creating prosperity and deepening participation in the world's networks of exchange and production (Castells, 1996).

Even if the world remains short of truly universal access, the mobile seems to be more intertwined with our identities as human beings. In 1954, Harold Osborn suggested that, at birth, people would soon be given a number that eventually would be their phone number. In an era that clearly pre-dated caller ID, and in an era which assumed that we gladly accepted any call that came our way, he noted morbidly that if someone tried to call you and you did not answer, they would know you were dead (Ling, 2004). The idea that there would be universal coverage via mobile communication was an engineer's dream associated with the development of the transistor. In an interesting echo of Osborn's notion of personal ID *qua* telephone number, Araba Sey has described how people in Ghana often write their mobile phone numbers on the entry to their homes as a type of address (Sey, 2007). If they are not home, the frustrated visitor knows how to get in touch. The same concept was taken to a humorous extreme in a recent advertising campaign for a mobile operator in India. In the advertisements, a famous Bollywood celebrity was portrayed as the wise leader of a village rife with community tension. His idea was to replace everyone's names (loaded as they are with markers of social status and affiliations) with their mobile numbers; happiness and peace ensues.[11] It has been a responsibility and privilege of the state in the modern era to assign and maintain records and markers of human identities, birth certificates, last names, addresses, and so on (Scott, 1998). It is remarkable to see how these are augmented by, and in some cases replaced by, the power of a single string of digits, issued by a mobile company, and carried with us for life. And it is not surprising that many consumers assume that this identity is theirs, and should be "portable" (transferrable from one carrier to another) even though the numbers are assigned and maintained by the carriers (Sang-Woo, Kim & Myeong-Cheol, 2004).

5 Mobile convergence: is the mobile phone a Swiss Army knife?

As we near the close of this volume, it should be clear that we have spent the bulk of our analysis on two large themes of basic mediated/mobile communication: the stunning rise in the access to telephony around the world, and the possibilities for new forms of social ties, organizations and behaviors brought about by individual addressability and the interlacing of individual activities. Far from drawing on the "Swiss Army knife" do-it-all handsets and 3G networks underpinning "convergence culture" (Jenkins, 2006), the examples we have used to illustrate these themes have focused primarily on voice and simple text messaging. This is not to say that mobiles are not already being integrated into the realm of converged media and portable computing. Hundreds of millions of people share content, ringtones and pictures across phones. A billion or more mobile cameras capture life's trivialities, joys and tragedies. With every passing day, smarter, more powerful handsets and networks let us do more with our time: listen to music, play games, shoot photos and video, gather important financial and health information, organize our day, check e-mail, and more easily access the internet and the World Wide Web. Yet for every celebrity scandal captured on a mobile video camera, there are literally millions of calls to coordinate picking up the kids or simple "how are you?" text messages. For every Alberto checking e-mail on the go and every Stan checking Facebook mobile, there are Amihans and Mettes and Rohits who basically use the mobile as a phone. To lose track of this fact is to lose track of the social impact of the mobile phone *so far*. It is remarkable, in a way, how mobile use often remains so mundane.

So, we've focused, through vignettes and examples, on the everyday, perhaps to the relative exclusion of an exploration of the role mobiles will play in the emerging convergence culture (Jenkins, 2006), or the impact super-networked, mobile media will play in the next decade and beyond (Goggin, 2006). We know, however, that these changes are both inevitable and unpredictable. Regional differences will emerge and cross-pollinate, from QQ's mobile instant messenger application in China to Facebook's IM application in the USA. New stresses will emerge as users negotiate how to integrate ever-more-powerful mobile devices into their lives. But the question of whether the mobile internet will reach a majority of the world's mobile users, beyond those billion or so who use the traditional internet, remains open. Even if the voice gap disappears, a data gap may remain.

At this time, our focus on the way the mobile has been adopted, domesticated and appropriated as a phone by billions of users can help sensitize us to the questions that lie ahead.

At the risk of setting up too stark a dichotomy, we could say that our analysis of the pattern of voice calls has illustrated that, while mobile communication can certainly be used to convey critical information (on crop prices, on health care, about emergencies, about transit delays, etc.), to reduce the technology's utility to the transmission of information is to overlook its role in connecting loved ones, expressing identities and entertaining individuals. (Again, we are re-visiting arguments made by Fischer [1992], about landlines.) As more mobiles become better conduits for non-voice data, whether via the mobile internet, or video or even the humble SMS, the same dichotomies will persist; these new data channels will convey and process valuable information *and* will enable self-expression, social connection and entertainment, as well as various forms of mischief, in new and unanticipated ways.

While it is tempting to view the arrival of a universally adopted, low-cost means to access the internet as revolutionary, the perspectives offered in this book might allow us to interpret this arrival in more evolutionary terms. The two primary lenses we have established (that mobiles extend connectivity and connection to the previously unconnected, and that mobiles complicate time and space norms, enable interlacing and individual addressability), may go a long way towards predicting how mobile data use becomes a part of everyday life, relative to conventional voice telecommunications and to PC-based internet access.

6 Conclusion

Moving from the situation described by the Maitland Report, where the preponderance of people in the world did not have access to telephony, there has been an extensive transition. The technology for mobile communication has matured. At the time of the Maitland Report, mobile phones were barely out of the laboratories and they were ponderous, heavy and cumbersome. They were often owned only by people who were excessively rich.[12] Conditions have changed quickly; a virtuous cycle of innovation, investment and a broader customer base has brought handset prices down. Similar progress on the network side, plus competitive markets and innovations in billing like the prepay plan, have lowered tariffs and extended access to an ever higher proportion of the world's population. The effect of these lower prices has been an explosion in adoption that has surpassed even the most bullish of early predictions. In developed countries, a basic mobile phone is easily accessible for the majority of people. In developing countries, while it is not the first thing that an impoverished person uses money for (Krishna, 2004), it is something that is within reach of many people. The early chapters of this book described the two great transformations in telecommunications brought about by mobiles: in the developing world, mobiles have brought the telecommunications network within the reach of hundreds of millions, perhaps billions, of people; in the developed world, mobiles are achieving near-ubiquity, with some societies having far more mobile subscriptions than people. As a result, people are simply easier to reach than in the days of desk phones and phone booths.

What is the effect of this on society? The latter chapters of the book have outlined two of the primary effects of these transitions: the extension of the reach of connectivity across the planet, and the intensification of the reachability of individuals – as opposed to places. A higher proportion of the world's population is engaging in a finer-grained interlacing of activities than was previously possible. We are micro-coordinating and we are flirting/gossiping/joking with our near friends and our families. We are organizing pick-ups for our children and we are finding the prices of commodities in what approximates a real time. Lovers are coordinating their tête-à-têtes, buddies are organizing their card games, teens are organizing their parties and thieves are organizing their nocturnal activities. The interlacing of these interactions and the social dealings that are a part of them are resulting in tighter social bonds (for both good and ill). We are coming to rely on mobile communication and, indeed, are developing a mobile logic, which restructures the way we deal with the exigencies of everyday life. While "clock time" is still an essential basis for coordination, mobile communication is softening our schedules and allowing us to work out the best time and place to meet with our friends and family, or to find work and pursue our livelihoods.

The effects of the mobile phone are being felt around the globe. The teen in LA, the soccer mom in Oslo, the retirees in Chile, the restaurant owner in Rwanda and the migrant worker in China all own one. The rise of cheap handsets, inexpensive subscriptions and innovative forms of use will fan the further diffusion of mobile communication. It has established a logic of use that is difficult to ignore. It can be disruptive. It can jigger the way power is exercised and it can facilitate unwanted disruptions in our lives. It can stress us out and it can also save the day. Despite the debates and the tensions it creates, most of us crave it – either a first one or a new one. Indeed, most of us can no longer imagine life without it; this very unimaginability is the best indication of its impact on our lives, and the best indication that it has already proved to be a revolutionary device.

Notes

I INTRODUCTION: THE QUARTER-CENTURY
BEYOND THE MAITLAND COMMISSION REPORT

1 In this book we use the terms "mobile telephony," "mobile telephone," etc., to refer to what people in North America call "cellular telephony" or "cell phones."

2 SIM cards are the tiny, removable "Subscriber Identity Modules" which contain the account information for users on the GSM mobile networks. They can be swapped from handset to handset.

3 This telephone survey included 1,000 randomly selected persons aged 13 or older. It was carried out in June 2006 by Telenor.

4 Approximately 80 percent of 10-year-old children also had a mobile telephone.

5 Telephone ladies run small businesses in the developing world, using mobile phones like payphones for their communities (Aminuzzaman et al., 2003). They receive a "micro-loan" from, for example, the Grameen Bank. With that money, they buy a mobile phone and, if necessary, a solar battery recharger or extended-range antenna. The telephone ladies use the receipts from these calls to pay off their loan and the rest becomes a contribution to the household economy. Umbrella ladies do something similar, but without the formal franchise affiliations and micro-loans. Of course, men run these businesses as well, but it is the ladies who have entered the popular lore and lexicon.

2 SHORT HISTORY OF MOBILE COMMUNICATION

1 With the rise of internet telephony and indeed mobile internet telephony, it is not just the networks of the telecom operators that mediate our conversations and our text messages.

2 The number of radio cells needed to cover a particular area depends on its physical topography – areas affording line of sight

are easier to cover than rolling terrain – the presence or absence of barriers such as reinforced concrete and the characteristics of the mobile handset and its antenna. A typical cell might be 1 kilometer from side to side. In some cases, it might be smaller and in areas with little use it might be larger.

3 Which is in turn based on telegraphy (Standage, 1998).

4 The exchange allows the critical function of routing calls from and to any telephone in the system (Cherry, 1977)

5 The first use by police came in Detroit in 1921. In this case it only included one-way communication. Upon receiving a message, the officer would have to stop his car to find a telephone in order to call back into headquarters (Agar, 2003).

6 A central part of the system was the development of sophisticated protocols for sharing radio frequencies.

7 For example, the auctioning of the 700MHz range given up by television broadcasters in favor of digital transmission could raise as much as $15 billion for the federal treasury of the US government.

8 Almost as if to underscore the interest in mobile communication, the 1980s saw the rise of Citizens' Band radio. This was/is a form of broadcast radio that does not afford the privacy of telephony. In addition, there was an avid interest in paging systems that lasted until the early 1990s when it was replaced by enthusiasm for mobile communication. The use of these technologies illustrates the interest in mobile communication at the time (Goggin, 2006).

9 In June 2007 there were 2 billion GSM users and a total of 2.29 billion mobile telephone users of all types. Thus about 87 percent of all mobile phones are variations of the GSM standard. More correctly this should be referred to as the GSM group of technologies, which includes, among others, the following family of technologies, in order of increasing mobile band width: Global System for Mobile Communications (GSM), General Packet Radio Service (GPRS), Enhanced Data rates for GSM Evolution (EDGE), Universal Mobile Telecommunications System (UMTS) and High-Speed Downlink Packet Access (HSDPA).

10 In addition to the iPhone, the Japanese operator KDDI has a system called EZweb, and J-Phone has a system called J-Sky.

11 Or, more correctly, mobile terminals.

12 Ironically, the development of more advanced handsets that include, for example, WiFi, Bluetooth and Global Positioning,

color displays, flash devices for their cameras, etc., cuts in the opposite direction.

13 A SIM is a small chip attached to a piece of plastic. The whole card is slightly larger than a grown person's thumbnail. In GSM telephones, the SIM card is inserted into the mobile phone. It is where the "subscription" resides. Thus, rather than buying a handset from the operator, GSM allows the user to buy the subscription separately from the actual handset. Thus, the user can change telephones by simply removing the SIM card and placing it into another telephone handset.

14 See www.janchipchase.com/blog/mt-search.cgi?tag=used&blog_id=1.

15 There are also examples of prepayment for other utilities such as electricity and gas. Prepaid mobile telephony was first trialed in Mexico in 1992. Somewhat later it was adopted in Portugal and Italy, and since has become the most common form of mobile subscription (Kalba, 2008, 642).

16 An exception is when the person receiving the call is "roaming" outside the coverage area of their national operator. In that case, the person making the call pays the normal price for the call, but the person receiving the call pays for the delivery of the call from the home country to the country where they are.

17 There are other dimensions of calling in the USA that ameliorate the cost of use. Among other things, there are no long-distance fees with mobile phone use. Given the size of the USA and the mobility of people, this can represent a considerable saving. In addition, there is often free calling in the evening and at weekends, and there is calling within "family plans" that often does not count against the minutes in a calling plan. Finally, unused "minutes" from one month can be "rolled over" to the next month. (Thanks to Leysia Palen for this information.)

18 In some cases, as with the iPhone, the operators also agree that they will pay the handset producers a portion of the revenue that the handset generates.

3 MOBILE COMMUNICATION IN EVERYDAY LIFE:
3 BILLION NEW TELEPHONES

1 Intentional missed calls are known by many names: "beeps," "flashes," "pranks," *llamadas perdidas*, etc. The goal of a missed

call is to ring a person's mobile once, thus leaving their number via the caller ID system, and hang up before the call is complete; a message is conveyed without either party paying for an SMS or voice call. Generally, if the receiver of the beep recognizes the number of the person who is beeping her, or if the receiver has programmed the number of the beeper in her mobile's address book, then the identity of the beeper is clear. It follows that a 'beep' can mean a wide variety of things, depending on the relationship between the receiver and the beeper, and the context of the moment. Most beeps mean "call me back." Others convey a pre-negotiated message, like "pick me up now," or, in Annette's case, "lunch is ready." Still others just mean "I miss you" or "I'm thinking of you." The practice of missed calls predates mobile phones, but it has exploded in popularity, particularly in the developing world where a combination of factors (calling-party-pays plans, reliance on prepaid airtime, and resource-constrained users) makes it attractive. Some estimates suggest that missed calls may account for 20 to 30 percent of all initiated calls on some networks in Africa and South Asia (Donner, 2007).

2 Readers are encouraged to refer to a collection of papers in *The Social Impact of the Telephone* on this topic (de Sola Pool, 1977).

4 MOBILE COMMUNICATION IN EVERYDAY LIFE: NEW CHOICES, NEW CHALLENGES

1 Interestingly, the males in the survey had also adopted this ruse. Their motivation, however, was to avoid talking to persons with whom they did not want to speak.

2 Bob Brotchie, a paramedic in the UK, suggested the system. It is now a standard type of entry for many mobile phone users.

3 www.textually.org/textually/archives/2007/10/017735.htm.

4 www.iht.com/articles/2008/02/20/technology/wireless21.php.

5 This has been the case with many types of technologies, ranging from automobiles to TVs to landline telephones to iPods.

6 This type of cycling between handsets is facilitated by the GSM SIM.

7 Even under the same roof, kids can be in their own place using their own telephone line (Pertierra, 2006; Pertierra, Ugarte, Pingol, Hernandez, & Dacanay, 2002).

5 DEBATES SURROUNDING MOBILE COMMUNICATION

1 There are those who suggest that in fact his wife did not call and that the whole episode was staged so as to capture the attention of the press and also to humanize Giuliani (Dobbs, 2007).

2 www.textually.org/textually/archives/2006/06/012544.htm.

3 This point of behavior has changed. Writing in the late 1950s, Goffman described the need to protect telephone conversations from being heard by others (1959, 186).

4 Writing before the rise of mobile telephony, Goffman described the "unboothed street phone" as presenting special social forms of interaction (1981, 86).

5 According to Rafael (2003), the protests against Estrada were more intense among the middle class than among the poorer sectors of the population. Once assembled, the protesters were moved – both physically and emotionally – with the more traditional tools of protest such as speeches, etc.

6 news.bbc.co.uk/1/hi/world/asia-pacific/7021402.stm.

7 It was reported that, during the visit of President George W. Bush, Australian authorities jammed phones in the areas he was visiting (McPhedran, 2007).

8 www.trapster.com/rl/Trapster.php.

9 www.nyscba.com/drunkstoppers.html.

10 www.cbsnews.com/stories/2007/01/02/iraq/main2320313.shtml.

11 It has been suggested that there is a 90:9:1 ratio of lurkers : occasional contributors : active users (Nielsen, 2006). Only a few users actively feed information into the system and the vast majority of users follow along in either some type of active lurking or bemused disengagement. On more popular sites the ratio can be even more skewed, to the degree that only a small per mille of users are also contributors. Thus, it is possible to ask if this is a type of "broadcasting" in that a small core of individuals produces content for the consumption of another broad group.

12 The survey, carried out by Telenor, was conducted in the spring of 1999 and included 1,006 people aged 13 to 20 years.

13 The material reported here is based on an online study, and thus it is difficult to generalize the results. The sample was recruited from a web site of a particular rock musician and, thus, while the analysis and research were carried out in the USA, the respondents could potentially come from anywhere on the globe. These percentages are generally the same levels as reported in

Norway by Tikkanen and Junge (2004). Another version of this phenomenon is so-called "happy slapping," in which a victim is attacked while others video the act and broadcast it (Brough & Sills, 2006). Thankfully, this appears to be more of a fad than a trend.

14 This analysis was based on survey data gathered from a random selection of 11,928 teens in Norwegian Statistical Research's study "Ung i Norge" (Young in Norway) collected in February of 2002.

15 www.earthtimes.org/articles/show/80873.html.

16 www.nytimes.com/2008/03/04/world/asia/04briefs-cell.html?_r =1&scp=2&sq=taliban&st=nyt&oref=slogin.

17 Taliban threatens Afghan cellphone companies. www.nytimes. com/2008/02/26/world/asia/26afghan.html?ref=todayspaper.

18 North Korea to allow mobile phones in Pyongyang. http://english.donga.com/srv/service.php3?bicode=060000 &biid=2008021912358. Cuba lifts mobile phone restrictions. www.guardian.co.uk/world/2008/mar/28/cuba?gusrc=rss&feed =networkfront.

6 CONCLUSION: INDIVIDUAL ADDRESSABILITY, INTERLACING AND THE SPILLOVER OF THE CONTROL REVOLUTION

1 This is not a universal truth. In many cases, the mobile telephone is considered a personal device. In other situations, such as in the developing world, it is more like a traditional house telephone where it is the property of a household or a collection of people.

2 The mobile telephone as a device with shared ownership can sometimes characterize an earlier phase in the adoption process. In Norway, for example, in 1999, 22.6 percent of the respondents in a nationally representative sample indicated that they shared a mobile phone with others in their household. At that time, 58 percent of the respondents reported that they personally owned their own device (n = 1,898). In 2006, 93.7 percent reported personally owing a device and only 3.9 percent reported that they shared one (n = 1,833) (Vaage, 2006).

3 It can also allow others to wedge their phone calls into time slots where we are already occupied – as discussed above. Thus we need to sometimes be quick on our feet in order to not unduly disturb the co-present situation. Thus, while there is an efficiency

associated with the ability to interlace interactions, it can also
result in various types of disruption.

4 The internet has become a medium through which much
commerce is carried out. It is also the locus of many forms of
social interaction, some more absurd than others. There are
social networking sites, instant message applications, e-mail.
There are locations used to search for old high-school friends,
archives of video material that range from footage of Roosevelt,
Churchill and Stalin at Yalta, to interviews with Jack Kerouac
and Allen Ginsberg, to sophomoric dorm-room productions and
a wealth of other material that has little direct connection to the
control of production processes.

5 An interesting contrast here is that intentional missed calls
are somewhat one-dimensional in this respect. Where iterative
planning progressively works out an agreement, missed calls
assume that both partners have already agreed on the meaning of
the missed call and there is no need for it to be expanded upon.
That is, there is no need for iterations in the interaction. (They
can have a more expressive side when, for example, the enamored
lover sends 50 missed calls to the object of his/her desires
[Geirbo et al., 2007]). In this way, there is the direct addressability
and there is the interlacing into the flow of daily life. However,
there is not the burst of interactions. An underlying message
is that iterative planning is a luxury of those operating without
economic constraints.

6 It is clear that the mobile phone has many other features. It can
be a camera, a device for listening to music, a calendar and a
Dictaphone. In addition, it is possible to configure the mobile
phone to include functions such as banking transactions,
blogging, watching TV, connecting a PC to the net, location-based
services, video telephony, IM, gaming, social networking, reading
e-mail, etc. While there is a wide functionality associated with the
mobile phone, at its core it is still a communications terminal
that we use in order to talk (or text) with others.

7 It is instructive to place the use of the mobile phone in the
Tönnies *Gemeinschaft–Gesellschaft* continuum (Tonnies,
1963). He describes the general drift in society as going from
Gemeinschaft – that is, association based on shared belief, shared
place or shared kinship – towards *Gesellschaft*, where individual
self-interest is a more central driving force. On the one hand,
mobile communication can be seen as a part of the general drift
from *Gemeinschaft* towards *Gesellschaft*. Mobile communication

is a part of the general modernization of society and it arises from technical drift in society. In this phase of its use, it helps with improved efficiency and plays into economic development. It supports an increased connectivity to (and coordination with) international economic activity. However, it is also possible to say that it is a part of a re-emphasis on the *Gemeinschaft*. We are largely calling the core members of our intimate sphere and we are engaged in the Goffmanian everyday rituals that help to support social cohesion in these small intimate circles (Ling, 2008). We are engaged in interaction with our best friends and our family. We are creating proto-villages against the backdrop of industrialized society.

8 These statistics come from a series of surveys carried out by Telenor and organized by the first author. They all involve random samples of approximately 1,000 people in Norway aged 13 years and older. The surveys were carried out via telephone.

9 www.news.com/Cell-phone-coverage-holes-hurt-public-safety/2100-1039-3-6143866.html.

10 www.abc.net.au/news/stories/2008/03/06/2181800. htm?site=riverina.

11 http://economictimes.indiatimes.com/News/News_By_Industry/ Services/Advertising/Junior_Bs_new_Idea_TV_ad_gets_ political_colour/rssarticleshow/2437936.cms.

12 An encapsulation of this is seen in the movie *Wall Street* when the rich and ruthless financier Gordon Gekko uses an early mobile phone to call his protégé Bud Fox.

References

ABI Research. (2007). *Handset Recycling and Refurbishment*. Oyster Bay, NY: ABI Research.

Agar, J. (2003). *Constant Touch: A Global History of the Mobile Phone*. Cambridge: Icon Books.

Allen, S. (2007). Citizen journalism and the rise of "Mass self-communication": reporting the London bombings. *Global Media Journal* 1(1). Retrieved January 15, 2009. http://stc.uws.edu.au/gmjau/iss1_2007/pdf/HC_FINAL_Stuart%20Allan.pdf.

Aminuzzaman, S., Baldersheim, H., & Jamil, I. (2003). Talking back: empowerment and mobile phones in rural Bangladesh: a study of the village pay phone of Grameen Bank. *Contemporary South Asia* 12(3), 327–48.

Andrewes, W. J. H. (2002). A chronicle of timekeeping. *Scientific American* 287(3), 58–67.

Arnold, M. (2003). On the phenomenology of technology: the "Janus-faces" of mobile phones. *Information and Organization* 13(4), 231–56.

Aronson, S. H. (1977). Bell's electrical toy: what's the use? The sociology of early telephone usage. In I. de Sola Pool (ed.), *The Social Impact of the Telephone* (pp. 15–39). Cambridge, MA: MIT Press.

Bakken, F. (2005). SMS use among deaf teens and young adults in Norway. In R. Harper, L. Palen & A. Taylor (eds.), *The Inside Text: Social, Cultural and Design Perspectives on SMS* (pp. 161–73). Dordrecht, Netherlands: Springer.

Baldwin, D. (1997). "The Front Page" on speed (a behind-the-scenes look at media coverage of the death of Diana, Princess of Wales). *American Journalism Review* 19(8), 10–12.

Baron, N. (2008). *Always On: Language in an Online and Mobile World*. Oxford: Oxford University Press.

Baron, N., & Ling, R. (2007). Emerging patterns of American mobile phone use: electronically-mediated communication in transition. Paper presented at the Mobile Media conference, July 2–4, Sydney, Australia.

Basel Action Network (2004). Mobile toxic waste. *Journal.* Retrieved January 15, 2009 from www.ban.org/Library/mobilephonetoxicity rep.pdf.

Bastiansen, H. G. (2006). *Det piper og synger overalt: mobiltelefonen i Norge fra ide til allemannseie.* Oslo: Norsk Telemuseum.

Batista, E. (2001). SMS provides SOS lifeline. *Wired.* Retrieved January 15, 2009 from www.wired.com/science/discoveries/ news/2001/02/41621.

Bayes, A. (2001). Infrastructure and rural development: insights from a Grameen Bank village phone initiative in Bangladesh. *Agricultural Economics 25*(2–3), 261–72.

Beardsley, S., von Morgenstern, I. B., Enriquez, L., & Kipping, C. (2002). The elements of successful telecommunications sector reform. In G. Kirkman, P. Cornelius, J. Sachs & K. Schwab (eds.), *The Global Information Technology Report 2001–2002: Readiness for the Networked World* (pp. 138–58). New York: Oxford University Press.

Beniger, J. R. (1986). *The Control Revolution: Technological and Economic Origins of the Information Society.* Cambridge, MA: Harvard University Press.

Best, M. L. (2003). The wireless revolution and universal access. In ITU (ed.), *Trends in Telecommunication Reform* (pp. 107–22). Geneva: ITU.

Bijker, W. E., Hughes, T. P., & Pinch, T. (1987). *The Social Construction of Technological Systems: New Directions in the Sociology and Technology of History.* Cambridge, MA: MIT Press.

Blaise, C. (2000). *Time Lord: Sir Sanford Fleming and the Creation of Standard Time.* New York: Vintage.

Blumberg, S. J., & Luke, J. V. (2007). *Wireless Substitution: Early Release of Estimates From the National Health Interview Survey, January–June 2007.* Atlanta, GA: National Center for Health Statistics.

Boettinger, H. M. (1977). Our sixth-and-a-half sense. In I. de Sola Pool (ed.), *The Social Impact of the Telephone* (pp. 200–7). Cambridge, MA: MIT Press.

Brandtzæg, P. (2005). Children's use of communications technologies. Paper presented at the Mobile Media, Mobile Youth conference, September 30, 2005, Copenhagen.

Brooks, J. (1976). *Telephone: The First Hundred Years.* New York: Harper and Row, Publishers.

Brough, R., & Sills, J. (2006). Multimedia bullying using a website. *British Medical Journal 91*(2), 202.

Bull, M. (2001). The world according to sound: investigating the world of Walkman users. *New Media Society 3*(2), 179–97.

Burgess, A. (2004). *Cellular Phones, Public Fears, and a Culture of Precaution.* Cambridge: Cambridge University Press.

Burns, P. C., Parkes, A., Burton, S., Smith, R. K., & Burch, D. (2002). *How Dangerous is Driving with a Mobile Phone? Benchmarking the Impairment of Alcohol.* Crowthorne, UK: Transport Research Laboratory.

Cain, A., & Burris, M. (1999). Investigation of the use of mobile phones while driving. Retrieved March 5, 2003, from www.cutr.usf.edu/its/mobile_phone.htm.

Campbell, S. W. (2006). Perceptions of mobile phones in college classrooms: ringing, cheating, and classroom policies. *Communication Education* 55(3), 280–94.

Carey, J. W. (1988). *Communication as Culture: Essays on Media and Society.* New York: Routledge.

Carlsson, Y., & Decker, S. H. (2005). Gang and youth violence prevention and intervention: contrasting the experience of the Scandinavian welfare state with the United States. In S. H. Decker & F. M. Weerman (eds.), *European Street Gangs and Troublesome Youth Groups* (pp. 259–86). Lanham, MD: Rowman Altamira.

Carroll, C. (2008). High-tech trash. *National Geographic* 213(January), 64–81.

Cartier, C., Castells, M., & Qiu, J. L. (2005). The information haveless: inequality, mobility, and translocal networks in Chinese cities. *Studies in Comparative International Development* 40(2), 9–34.

Casas, C., & LaJoie, W. (2003). Voxiva: Peru. Retrieved September 30, 2004, from www.bus.umich.edu/BottomOfThePyramid/Voxiva.pdf.

Castells, M. (1996). *The Rise of the Network Society* (Vol. I). Malden, MA: Blackwell.

Castells, M., Fernández-Ardèvol, M., Qiu, J. L., & Sey, A. (2007). *Mobile Communication and Society: A Global Perspective (Information Revolution and Global Politics).* Cambridge, MA: MIT Press.

Cellular News. (2008). Mobile phone subscribers pass 4 billion mark. Retrieved January 9, 2009, from www.cellular-news.com/story/35298.php.

Center for Migrant Advocacy. (2006). SOS SMS for overseas Filipino workers in distress. *Kasama* 20(2).

Chalfen, R. (1987). *Snapshot Versions of Life.* Bowling Green, OH: Bowling Green State University Popular Press.

Chavan, A. L. (2007). A dramatic day in the life of a shared Indian mobile phone. In N. M. Aykin (ed.), *Usability and Internationalization. HCI and Culture, Proceedings of the Second International Conference on Usability and Internationalization, UI-HCII 2007, Beijing, China, July 22–27* (pp. 19–26). New York: Springer.

Cherry, C. (1977). The telephone system: creator of mobility and social change. In I. de Sola Pool (ed.), *The Social Impact of the Telephone* (pp. 112–26). Cambridge, MA: MIT Press.

Chesley, N. (2005). Blurring boundaries? Linking technology use, spillover, individual distress, and family satisfaction. *Journal of Marriage and Family* 67, 1237–48.

Cohen, A. A., & Lemish, D. (2005). When bombs go off the mobiles ring: the aftermath of terrorist attacks. In K. Nyíri (ed.), *A Sense of Place: The Global and the Local in Mobile Communication* (pp. 117–28). Vienna: Passagen Verlag.

Cohen, A. A., Lemish, D., & Schejter, A. M. (2007). *The Wonder Phone in the Land of Miracles: Mobile Telephony in Israel*. Cresskill, NJ: Hampton Press.

Collins, R. (2004). *Interaction Ritual Chains*. Princeton, NJ: Princeton University Press.

Cottrell, W. F. (1945). Death by dieselization: a case study in the reaction to technological change. *American Sociological Review* 16, 63–75.

Crawford, M. (1994). The world in a shopping mall. In M. Sorkin (ed.), *Variations on a Theme Park: The End of Public Space* (pp. 3–30). New York: Hill and Wang.

Cumming-Bruce, N. (2005). Wireless: in Thailand, on the trail of cellphone terrorists. *International Herald Tribune*, May 2, from www.iht.com/articles/2005/05/01/business/wireless02.php.

de Sola Pool, I. (ed.). (1977). *The Social Impact of the Telephone*. Cambridge, MA: MIT Press.

Dobbs, M. (2007). Rudy's "spontaneous" cell phone "stunt". *Washington Post*, October 8, from http://voices.washingtonpost.com/fact-checker/2007/10/let_rudy_be_rudy.html.

Donner, J. (2004a). Innovative approaches to public health information systems in developing countries: an example from Rwanda. Paper presented at the Mobile Technology and Health: Benefits and Risks conference, June 7–8, University of Udine Department of Economics, Society, and Geography, Udine, Italy.

(2004b). Microentrepreneurs and mobiles: an exploration of the uses of mobile phones by small business owners in Rwanda. *Information Technologies and International Development* 2(1), 1–21.

(2005). The social and economic implications of mobile telephony in Rwanda: an ownership/access typology. In P. Glotz, S. Bertschi & C. Locke (eds.), *Thumb Culture: The Meaning of Mobile Phones for Society* (pp. 37–52). Bielefeld: Transcript Verlag.

(2007). The rules of beeping: exchanging messages via intentional "missed calls" on mobile phones. *Journal of Computer-Mediated*

Communication 13(1), 1–22.

(2008a). Research approaches to mobile use in the developing world: a review of the literature. *Information Society* 24(3), 140–59.

(2008b). Shrinking fourth world? Mobiles, development, and inclusion. In J. Katz (ed.), *Handbook of Mobile Communication Studies* (pp. 29–42). Cambridge, MA: MIT Press.

Donner, J., Rangaswamy, N., Steenson, M. W., & Wei, C. (2008). "Express Yourself" and "Stay Together": the middle-class Indian family. In J. Katz (ed.), *Handbook of Mobile Communication Studies* (pp. 325–37). Cambridge, MA: MIT Press.

Donner, J., & Tellez, C. A. (2008). Mobile banking and economic development: linking adoption, impact, and use. *Asian Journal of Communication* 18(4), 318–22.

Doussantousse, S., & Tooke, L. (2002). Women involved in prostitution in Viet Nam – a Hanoi snapshot. *Current Trends in Prostitution in Viet Nam*. Retrieved January 15, 2009 from www.unaids.org.vn/resource/topic/sexwork/CSWprostitution.doc.

Durkheim, E. (1995). *The Elementary Forms of Religious Life* (trans. K. E. Fields). Glencoe, IL: The Free Press.

Dutton, W. (2003). The social dynamics of wireless on September 11: reconfiguring access. In A. M. Noll (ed.), *Crisis Communication* (pp. 69–82). Lanham, MD: Rowan and Littlefield.

Dyckman, J. W. (1973). Transportation in the cities. In K. Davis (ed.), *Cities, Their Origin, Growth and Human Impact* (pp. 195–206). San Francisco: W. H. Freeman.

Dymond, A., & Oestmann, S. (2003). The role of sector reform in achieving universal access. In *Trends in Telecommunication Reform 2003* (pp. 51–64). Geneva: ITU.

Eggleston, K., Jensen, R., & Zeckhauser, R. (2002). Information and telecommunication technologies, markets, and economic development. In G. Kirkman, P. Cornelius, J. Sachs & K. Schwab (eds.), *The Global Information Technology Report 2001–2002: Readiness for the Networked World* (pp. 62–75). New York: Oxford University Press.

Ellwood-Clayton, B. (2003). Virtual strangers: young love and texting in the Filipino archipelago of cyberspace. In K. Nyíri (ed.), *Mobile Democracy: Essays on Society, Self, and Politics* (pp. 225–39). Vienna: Passagen Verlag.

(2005). Texting and God: the Lord is my textmate – folk Catholicism in the cyber Philippines. In K. Nyíri (ed.), *A Sense of Place: The Global and the Local in Mobile Communication* (pp. 251–65). Vienna: Passagen Verlag.

Engvall, A., & Hesselmark, O. (2004, October). Profitable Universal

Service Providers. Retrieved April 10, 2006, from www.eldis.org/fulltext/profitable.pdf.

Eriksen, T. H. (2001). *Øyeblikkets tyranni: rask og langsom tid i informasjonssamfunnet* (Tyranny of the moment: fast and slow time in the information age). Oslo: Aschehoug.

Farley, T. (2005a). Mobile telephone history. *Telektronikk 3/4*(2005), 22–34.

(2005b). Privateline.com: telephone history. Retrieved January 15, 2009 from http://privateline.com/PCS/history.htm.

Fischer, C. S. (1992). *America Calling: A Social History of the Telephone to 1940*. Berkeley, CA: University of California Press.

Fischer, C. S., & Carroll, G. R. (1988). Telephone and automobile diffusion in the United States, 1902–1937. *American Journal of Sociology* 93(5), 1153–78.

Forestier, E., Grace, J., & Kenny, C. (2002). Can information and communication technologies be pro-poor? *Telecommunications Policy* 26(11), 623–46.

Fortunati, L. (2005a). The mobile phone as technological artifact. In P. Glotz, S. Bertschi & C. Locke (eds.), *Thumb Culture: The Meaning of Mobile Phones for Society* (pp. 149–60). Bielefeld: Transcript Verlag.

(2005b). Mobile phones and fashion in post-modernity. *Telektronikk* 3/4(2005).

(2005c). Mobile telephone and the presentation of self. In R. Ling & P. Pedersen (eds.), *Mobile Communications: Re-negotiation of the Social Sphere* (pp. 203–18). London: Springer.

Frey, S. R., Dietz, T., & Kalof, L. (1992). Characteristics of successful American protest groups: another look at Gamson's strategy of social protest. *American Journal of Sociology* 98(2), 368–87.

Furst, R. T., Herrmann, C., Leung, R., Galea, J., & Hunt, K. (2004). Heroin diffusion in the mid-Hudson region of New York State. *Addiction* 99(4), 431–41.

Galperin, H., & Girard, B. (2005). Microtelcos in Latin America. In H. Galperin & J. Mariscal (eds.), *Information Technology and Poverty Alleviation: Perspectives from Latin America and the Caribbean* (pp. 93–115). Ottawa: IDRC.

Gamson, W. (1975). *The Strategy of Social Protest*. Chicago: Dorsey.

Garfinkel, H. (1967). *Studies in Ethnomethodology*. New York: Basic.

Gaver, W. W. (1991). Technology affordances. Paper presented at the CHI '91 conference, April 28 – May 2, New Orleans.

Geirbo, H. C., Helmersen, P., & Engø-Monsen, K. (2007). *Missed Call: Messaging for the Masses. A Study of Missed Call Signalling Behavior in Dhaka*. (Internal Telenor R&I publication.) Fomebu: Telenor R&I.

Adnan Ahmer Qayyum
1504372085   FC   05/01/2015   M   100709565-00
70 Robinson Ave   Ottawa   05/01/1969
(613) 23-   2 16 02.20:00 A   K1N8N9
Dr. S. Chartrand

Impetigo

Adnan Ahmer Qayyum
1504372085   FC   05/01/2015   M   100709565-00
70 Robinson Ave   Ottawa   05/01/1969
(613) 233-7773   K1N8N9
Dr. S. Chartrand   02/06/2010   10:20:00 A

Gibsen, J. J. (1979). *The Ecological Approach to Visual Perception*. New York: Houghton Mifflin.

Goffman, E. (1959). *The Presentation of Self in Everyday Life*. New York: Doubleday Anchor Books.

(1967). *Interaction Ritual: Essays on Face-to-Face Behavior*. New York: Pantheon.

(1981). *Forms of Talk*. Philadelphia: University of Pennsylvania Press.

Goggin, G. (2006). *Cell Phone Culture: Mobile Technology in Everyday Life*. London: Routledge.

González, M. C., Hidalgo, C. A., & Barabási, A.-L. (2008). Understanding individual human mobility patterns. *Nature* 453(7196), 779.

Goodman, J. (2004). Return to vendor: how second-hand mobile phones improve access to telephone services. Retrieved July 15, 2005, from www.vodafone.com/etc/medialib/attachments/cr_downloads. Par.91807.File.dat/Return_to_vendor_2.pdf.

(2005). Linking mobile phone ownership and use to social capital in rural South Africa and Tanzania. Retrieved August 17, 2007, from www.vodafone.com/etc/medialib/attachments/cr_downloads. Par.78351.File.tmp/GPP_SIM_paper_3.pdf.

Gordon, J. (2007). The mobile phone and the public sphere: mobile phone usage in three critical situations. *Convergence* 13(3), 307–19.

Grace, J., Kenny, C., & Qiang, C. Z. W. (2004). *Information and Communication Technologies and Broad-Based Development: Partial Review of the Evidence*. Washington, DC: World Bank Publications.

Grahm, S., & Marvin, S. (1996). *Telecommunications and the City: Electronic Spaces and Urban Places*. London: Routledge.

Gray, V. (2005). *The Un-wired Continent: Africa's Mobile Success Story*. Geneva: ITU.

Gripsrud, M. (in progress). Kommunikasjonens tvillingpar. *Norsk Medietidsskrift*.

GSM Association. (2007). Global money transfer pilot uses mobile to benefit migrant workers and the unbanked. Retrieved May 10, 2007, from www.gsmworld.com/news/press_2007/press07_14.shtml.

(2008). 20 facts for 20 years of mobile communications. Retrieved March 8, 2008, from www.gsmtwenty.com/20facts.pdf.

Haddon, L. (2003). Domestication and mobile telephony. In J. E. Katz (ed.), *Machines that Become Us* (pp. 43–56). New Brunswick, NJ: Transaction.

(2004). *Information and Communication Technologies in Everyday Life*. Oxford: Berg.

(ed.). (1997). *Communications on the Move: The Experience of Mobile Telephony in the 1990s*. Farsta: Telia.

Hall, P. (1996). *Cities of Tomorrow: An Intellectual History of Urban Planning and Design in the 20th Century*. Oxford: Blackwell.

Hamilton, J. (2003). Are main lines and mobile phones substitutes or complements? Evidence from Africa. *Telecommunications Policy 27*, 109–33.

Hammond, A. L. (2007). *The Next 4 Billion: Market Size and Business Strategy at the Base of the Pyramid*. Washington, DC: World Resources Institute, International Finance Corporation.

Hård af Segerstad, Y. (2005). Language in SMS – a socio-linguistic view. In R. Harper, L. Palen & A. Taylor (eds.), *The Inside Text: Social, Cultural and Design Perspective on SMS* (pp. 33–52). Dordrecht, The Netherlands: Springer.

Hardy, A. (1980). The role of the telephone in economic development. *Telecommunications Policy 4(4)*, 278–86.

Harper, R., Palen, L., & Taylor, A. (eds.). (2005). *The Inside Text: Social, Cultural and Design Perspectives on SMS*. Dordrecht, The Netherlands: Springer:

Hills, J. (1990). The telecommunications rich and poor. *Third World Quarterly 12(2)*, 71–90.

Hjorthol, R., Jakobsen, M. H., Ling, R., & Nordbakke, S. (2007). Det mobile hverdagsliv: kommunikasjon og koordinering i moderne barnefamilier. In M. Lüders, L. Prøitz & T. Rasmussen (eds.), *Personlige medier. Livet mellom skjermene*. Oslo: Gyldendal Akademisk.

Hochschild, A. R. (2003). *The Time Bind: When Work Becomes Home and Home Becomes Work. The Cultural Study of Work*. New York: Henry Holt and Company.

Horst, H., & Miller, D. (2006). *The Cell Phone: An Anthropology of Communication*. Oxford: Berg.

Hudson, H. E. (1984). *When Telephones Reach the Village: The Role of Telecommunications in Rural Development*. Norwood, NJ: Ablex.
(2006). *From Rural Village to Global Village: Telecommunications for Development in the Information Age*. Mahwah, NJ: Lawrence Erlbaum Associates.

Hutchby, I. (2001). *Conversation and Technology*. Cambridge: Polity.

Ibarguen, G. (2003). Liberating the radio spectrum in Guatemala. *Telecommunications Policy 27(7)*, 543–54.

ICBC (Insurance Corporation of British Columbia). (2001). *The Impact of Auditory Tasks (as in Hand-Free Cell Phone Use) on Driving Task Performance*. North Vancouver, BC: ICBC Transportation Safety Research.. Retrieved January 15, 2009 from www.icbc.com/library/research_papers/cell_phones/images/cellphones_impact2.pdf#search=%22The%22.

Idowu, B., Ogunbodede, E., & Idowu, B. (2003). Information and communication technology in Nigeria: the health sector experience. *Journal of Information Technology Impact* 3(2), 69–76.

IE Market Research. (2008). 1Q08 mobile forecast: India, 2007–2010. Available from www.researchandmarkets.com/reports/c83235.

infoDEV. (2004). Case study: Manobi. Retrieved September 8, 2005, from www.sustainableicts.org/infodev/Manobi.pdf.

——— (2006). Micro-payment systems and their application to mobile networks. Retrieved May 27, 2008, from http://infodev.org/files/3014_file_infoDev.Report_m_Commerce_January.2006.pdf.

Ishii, K. (2006). Implications of mobility: the uses of personal communicatoion media in everyday life. *Journal of Communications* 56, 346–65.

Ito, M., Okabe, D., & Matsuda, M. (eds.). (2005). *Personal, Portable, Pedestrian: Mobile Phones in Japanese Life*. Cambridge, MA: MIT Press.

ITU. (2008a). Market information and statistics. Retrieved January 30, 2008, from www.itu.int/ITU-D/ict/statistics/.

——— (2008b). *Worldwide mobile cellular subscribers to reach 4 billion mark late 2008*. Retrieved January 15, from www.itu.int/newsroom/press_releases/2008/29.html.

Jagun, A., Heeks, R., & Whalley, J. (2007). Mobile telephony and developing country micro-enterprise: a Nigerian case study. Development Informatics Working Paper 29. Retrieved October 20, 2007, from www.sed.manchester.ac.uk/idpm/research/publications/wp/di/documents/di_wp29.pdf.

James, J., & Versteeg, M. (2007). Mobile phones in Africa: how much do we really know? *Social Indicators Research* 84(1), 117–26.

Jenkins, H. (2006). *Convergence Culture: Where Old and New Media Collide*. New York: New York University Press.

Jensen, R. (2007). The digital provide: information (technology), market performance and welfare in the South Indian fishers sector. *Quarterly Journal of Economics* 122(3), 879–924.

Julsrud, T., & Bakke, J. W. (2008). Trust, friendship and expertise: the use of email, mobile dialogues and SMS to develop and sustain social relations in a distributed work group. In R. Ling & S. W. Campbell (eds.), *The Mobile Communications Research Annual*, Vol. I: *The Reconstruction of Space and Time Through Mobile Communication Practices*. New Brunswick, NJ: Transaction.

Kalba, K. (2008). The adoption and diffusion of mobile phones – nearing the halfway mark. *International Journal of Communication* 2, 631–61.

Katz, J. E. (2005). Mobile phones in educational settings. In K. Nyíri (ed.), *A Sense of Place: The Global and the Local in Mobile Communication* (pp. 305–17). Vienna: Passagen Verlag.

——— (2006). *Magic in the Air: Mobile Communication and the Transformation of Social Life.* New Brunswick, NJ: Transaction.

Katz, J. E., & Aakhus, M. (2002). Conclusion: making meaning of mobiles – a theory of *Apparatgeist.* In J. E. Katz & M. Aakhus (eds.), *Perpetual Contact: Mobile Communication, Private Talk, Public Performance* (pp. 301–18). Cambridge: Cambridge University Press.

Katz, J. E., & Rice, R. E. (2002). The telephone as an instrument of faith, hope, terror and redemption: America, 9–11. *Prometheus* 20(3), 247–53.

Kavoori, A., & Chadha, K. (2006). The cell phone as a cultural technology: lessons from the Indian case. In A. Kavoori & N. Arceneaux (eds.), *The Cell Phone Reader: Essays in Social Transformation* (pp. 227–40). New York: Peter Lang.

Kidder, T. (1981). *The Soul of a New Machine.* Boston: Atlantic-Little, Brown.

Kim, H., Kim, G. J., Park, H. W., Sogang, A., & Nam, Y. (2006). The configurations of social relationships in communication channels: F2F, email, messenger, mobile phone, and SMS. Paper presented at the ICA Pre-conference on Mobile Communication, June 14–24, Erfert/Dresden, Germany.

King, J. L., & West, J. (2002). Ma Bell's orphan: US cellular telephony, 1947–1996. *Telecommunications Policy* 26(3–4), 189–203.

Konkka, K. (2003). Indian needs – cultural end-user research in Mumbai. In C. Lindholm, T. Keinonen & M. Spencer (eds.), *Mobile Usability: How Nokia Changed the Face of the Mobile Phone* (pp. 97–112). New York: Blackwell.

Koskinen, I. (2004). Seeing with mobile images: towards perpetual visual contact. In K. Nyíri (ed.), *A Sense of Place: The Global and the Local in Mobile Communication* (pp. 13–25). Vienna: Passagen Verlag.

Kowalski, R. M., & Linber, S. P. (2007). Electronic bullying among middle school students. *Journal of Adolescent Health* 41(6), S22–S30.

Krishna, A. (2004). Escaping poverty and becoming poor: who gains, who loses, and why? *World Development* 32(1), 121–36.

Kury, H., Chouaf, S., Obergfell-Fuchs, J., & Woessner, G. (2004). The scope of sexual victimization in Germany. *Journal of Interpersonal Violence 19*, 589–602.

Landes, D. S. (1983). *Revolution in Time: Clocks and the Making of the Modern World.* Cambridge, MA: Belknap Press.

Lasica, J. D. (2003). What is participatory journalism? *Online Journalism Review*. Retrieved January 15, 2009, from www.ojr.org/ojr/work-place/1060217106.php.

Law, P., & Peng, Y. (2007). Cellphones and the social lives of migrant workers in southern China. In R. Pertierra (ed.), *The Social Construction and Usage of Communication Technologies: Asian and European Experiences* (pp. 126–42). Quezon City: The University of the Philippines Press.

Leavitt, N. (2005). Mobile phones: the next frontier for hackers? *Computer 38*(4 April), 20–3.

Leland, J. (2005). Just a minute, boss. My cellphone is ringing. *New York Times*, Retrieved January 15, 2009 from www.nytimes.com/2005/07/07/fashion/thursdaystyles/07cell.html?_r=1.

Levinson, P. (2004). *Cellphone*. New York: Palgrave Macmillan.

Li, Q. (2007). New bottle but old wine: a research of cyberbullying in schools. *Computers in Human Behavior 23*(4), 1777–91.

Licoppe, C. (2004). Connected presence: the emergence of a new repertoire for managing social relationships in a changing communications technoscape. *Environment and Planning: Society and Space 22*, 135–56.

(2007). Co-proximity events: weaving mobility and technology into social encounters. Paper presented at the Towards a Philosophy of Telecommunications Convergence conference, September 27–29, Budapest.

Lindmark, S. (2002). Evolution of techno-economic systems: an investigation of the history of mobile communications. Unpublished doctoral dissertation, Chalmers University of Technology, Gothenberg, Sweden.

Ling, R. (2000). The impact of the mobile telephone on four established social institutions. Paper presented at the ISSEI2000 conference of the International Society for the Study of European Ideas, August 14–18, Bergen, Norway.

(2004). *The Mobile Connection: The Cell Phone's Impact on Society*. San Francisco: Morgan Kaufmann.

(2005a). Mobile communications vis-à-vis teen emancipation, peer group integration and deviance. In R. Harper, A. Taylor & L. Palen (eds.), *The Inside Text: Social Perspectives on SMS in the Mobile Age* (pp. 175–94). London: Kluwer.

(2005b). The socio-linguistics of SMS: an analysis of SMS use by a random sample of Norwegians. In R. Ling & P. Pedersen (eds.), *Mobile Communications: Renegotiation of the Social Sphere* (pp. 335–49). London: Springer.

(2007a). Children, youth and mobile communication. *Journal of Children and Media* 1(1), 60–7.

(2007b). The length of text messages and use of predictive texting: who uses it and how much do they have to say? In N. Baron (ed.), *Computer-Mediated Communication*. Washington, DC: AU TESOL Working Papers. Retrieved January 15, 2009, at www.american.edu/lfs/tesol/CMCLingFinal.pdf.

(2008). *New Tech, New Ties: How Mobile Communication is Reshaping Social Cohesion*. Cambridge, MA: MIT Press.

(2009). Mobile communication and teen emancipation. In G. Goggin & L. Hjorth (eds.), *Mobile Technologies: From Telecommunications to Media* (pp. 50–61). New York: Routledge.

Ling, R., & Baron, N. (2007). The mechanics of text messaging and instant messaging among American college students. *Journal of Sociolinguistics* 26(3), 291–8.

Ling, R., & Yttri, B. (2002). Hyper-coordination via mobile phones in Norway. In J. E. Katz & M. Aakhus (eds.), *Perpetual Contact: Mobile Communication, Private Talk, Public Performance* (pp. 139–69). Cambridge: Cambridge University Press.

(2006). Control, emancipation and status: the mobile telephone in teens' parental and peer relationships. In R. Kraut, M. Brynin & S. Kiesler (eds.), *Computers, Phones and the Internet: Domesticating Information Technology* (pp. 219–34). Oxford: Oxford University Press.

Ling, R., Yttri, B., Anderson, B., & DeDuchia, D. (2003). Mobile communication and social capital in Europe. In K. Nyíri (ed.), *Mobile Democracy: Essays on Society, Self and Politics* (pp. 359–74). Vienna: Passagen Verlag.

Lynne, A. (2000). *Nyansens makt – en studie av ungdom, identitet og klær* (Rapport 4 – 2000). Lysaker: Statens institutt for forbruksforskning.

Maitland, D. (1984). *The Missing Link: Report of the Independent Commission for Worldwide Telecommunications Development*. Geneva: International Telecommunication Union.

Mayer, M. (1977). The telephone and the uses of time. In I. de Sola Pool (ed.), *The Social Impact of the Telephone* (pp. 225–45). Cambridge, MA: MIT Press.

Mazzarella, W. (2003). *Shoveling Smoke: Advertising and Globalization in Contemporary India*. Durham, NC: Duke University Press.

McCombs, M. E., & Shaw, D. L. (1972). The agenda-setting function of mass media. *Public Opinion Quarterly* 36(2), 176–87.

McDowell, S. D., & Lee, J. (2003). India's experiments in mobile licensing. *Telecommunications Policy* 27(5–6), 371–82.

McKenzie-McLean, J. (2007). Abused women in fear of texts, emails. *Press*, October 8.

McPhedran, I. (2007). Bush shuts down Sydney trains. *Daily Telegraph*, from www.news.com.au/dailytelegraph/story/0,22049,21737457-5001021,00.html

Meyrowitz, J. (1985). *No Sense of Place: The Impact of Electronic Media on Social Behavior*. New York: Oxford University Press.

Miller, D. (2006). The unpredictable mobile phone. *BT Technology Journal 24*(3), 41–8.

Miyata, K. (2006). Longitudinal effects of mobile internet use on social network in Japan. Paper presented at the The International Communications Association conference, June 14–24, Dresden.

Molony, T. S. J. (2006). "I don't trust the phone; it always lies": trust and information and communication technologies in Tanzanian micro- and small enterprises. *Information Technologies and International Development 3*(4), 67–83.

Montague, D. (2002). Stolen goods: coltan and conflict in the Democratic Republic of Congo. *SAIS Review 22*(1), 103–18.

Mooallem, J. (2008). The afterlife of cell phones. *New York Times*, January 13.

Morawczynski, O., & Miscione, G. (2008). Examining trust in mobile banking transactions in Kenya: the case of M-PESA in Kenya. Paper presented at the IFIP WG 9.4-University of Pretoria Joint Workshop, September 23–24, Pretoria, South Africa.

Mumford, L. (1963). *Technics and Civilization*. San Diego: Harvest.

Nielsen, J. (2006). Participation inequality: encouraging more users to contribute. *Jakob Nielsen's Alertbox*, from www.useit.com/alertbox/participation_inequality.html.

Norman, D. (1990). *The Design of Everyday Things*. New York: Doubleday.

O'Connor, K. & Maher, C.A. (1982). Change in the spatial structure of a metropolitan region: work–residence relationships in Melbourne. In L. S. Bourne (ed.), *Internal Structure of the City: Readings from Urban Form, Growth and Policy* (pp. 406–21). New York: Oxford University Press.

Orlikowski, W. J., & Iacono, C. S. (2001). Research commentary: desperately seeking "IT" in IT research – a call to theorizing the IT artefact. *Information Systems Research 12*(2), 121.

Overå, R. (2008). Mobile traders and mobile phones in Ghana. In J. Katz (ed.), *Handbook of Mobile Communication Studies* (pp. 43–54). Cambridge, MA: MIT Press.

Pal, B. (2003). The doctor will text you now: is there a role for the mobile telephone in health care? *British Medical Journal* 326(7389), 607.

Palen, L., Salzman, M., & Youngs, E. (2001). Discovery and integration of mobile communications in everyday life. *Personal and Ubiquitous Computing* 5, 109–22.

Paragas, F. (2003). Dramatextism: mobile telephony and people power in the Philippines. In K. Nyíri (ed.), *Essays on Society, Self, and Politics* (pp. 259–83). Vienna: Passagen Verlag.

(2005). Migrant mobiles: cellular telephony, transnational spaces, and the Filipino diaspora. In K. Nyíri (ed.), *A Sense of Place: The Global and the Local in Mobile Communication* (pp. 241–9). Vienna: Passagen Verlag.

Patchin, J. W., & Hinduja, S. (2006). Bullies move beyond the schoolyard: a preliminary look at cyberbullying. *Youth Violence and Juvenile Justice* 4(2), 148–69.

Pedersen, P. (2005). Instrumentality challenged: the adoption of a mobile parking service. In R. Ling & P. Pedersen (eds.), *Mobile Communications: Re-negotiation of the Social Sphere* (pp. 369–72). London: Springer.

Pedersen, W., & Samuelsen, S. O. (2003). Nye mønstre av seksualatferd blant ungdom. *Tidsskrift for Den Norske Lægeforeningen* 21(6), 3006–9.

Pereira-Filho, J. L. (2003). Brazilian strategy on mobile spectrum. *Telecommunications Policy* 27(5–6), 333–50.

Pertierra, R. (2006). *Transforming Technologies: Altered States – Mobile Phone and Internet Use in the Philippines*. Manila: De La Salle University Press.

Pertierra, R., Ugarte, E. F., Pingol, A., Hernandez, J., & Dacanay, N. L. (2002). *Txt-ing selves: Cellphones and Philippine Modernity*. Manila: De La Salle University Press.

Pierce, J. R. (1977). The telephone and society in the past 100 years. In I. de Sola Pool (ed.), *The Social Impact of the Telephone* (pp. 159–95). Cambridge, MA: MIT Press.

Plant, S. (2002). *On the Mobile: The Effects of Mobile Telephones on Social and Individual Life*. Motorola. Retrieved January 15, 2009, from www.motorola.com/mot/doc/0/234_MotDoc.pdf.

Porteous, D. (2006). *The Enabling Environment for Mobile Banking in Africa*. London: DFID.

PT. (2008). *Det Norske Telemarked: 2007*. Oslo: Post- og Teletilsyn.

Qiu, J. L. (2007). The accidental accomplishment of Little Smart: understanding the emergence of a working-class ICT. *New Media & Society* 9(6), 903–23.

Rafael, V. L. (2003). The cell phone and the crowd: messianic politics in the contemporary Philippines. *Public Culture* 15(3), 399–425.

Rakow, L. (1992). *Gender on the Line: Women, the Telephone, and Community Life.* Champaign, IL: University of Illinois Press.

Rakow, L. F., & Navarro, V. (1993). Remote mothering and the parallel shift: women meet the cellular telephone. *Critical Studies in Mass Communication* 10, 144–57.

Raskauskas, J., & Stoltz, A. D. (2007). Involvement in traditional and electronic bullying among adolescents. *Developmental Psychology* 43(3), 564–75.

Recarte, M. A., & Nunes, L. M. (2000). Effects of verbal and spatial-imagery tasks on eye fixations while driving. *Journal of Experimental Psychology: Applied* 6(1), 31–43.

Redelmeier, D. A., & Tibshirani, R. J. (1997). Association between cellular-telephone calls and motor vehicle collisions. *New England Journal of Medicine* 336(7), 453–8.

Reid, D., & Reid, F. (2004). Insights into the social and psychological effects of SMS text messaging. *160 characters.* Retrieved January 15, 2009, from www.160characters.org/documents/SocialEffectsOfTextMessaging.pdf

Research and Markets. (2008). In 2008 it is predicted that over 2 trillion text messages will be sent worldwide and this number continues to grow. Retrieved January 15, 2009, from www.businesswire.com/portal/site/google/?ndmViewId=news_view&newsId=200802280 05080&newsLang=en.

Rheingold, H. (2002). *Smart Mobs.* Cambridge, MA: Persius.

(2003). Mobile virtual communities. *Journal 9.* Retrieved January 15, 2009, from http://66.102.1.104/scholar?num=50&hl=en&lr=&q=cache:LISoBrQ6ousJ:www.receiver.vodafone.com/06/articles/pdf/02.pdf+upoc.

Richtel, M. (2007). Devices enforce silence of cellphones, illegally. *New York Times*, November 4.

Rogers, E. (1995). *Diffusion of Innovations.* New York: The Free Press.

Roller, L.-H., & Waverman, L. (2001). Telecommunications infrastructure and economic development: a simultaneous approach. *American Economic Review* 91(4), 909–23.

Samuel, J., Shah, N., & Hadingham, W. (2005). Mobile communications in South Africa, Tanzania, and Egypt: results from community and business surveys. *Africa: The Impact of Mobile Phones.* Retrieved August 17, 2007, from www.vodafone.com/etc/medialib/attachments/cr_downloads.Par.78351.File.tmp/GPP_SIM_paper_3.pdf.

Sang-Woo, L., Kim, D. J., & Myeong-Cheol, P. (2004). Demand for number portability in the Korean mobile telecommunications market: contingent valuation approach. Paper presented at the 37th Annual Hawaii International Conference on System Sciences, January 5–8, 2004.

Saunders, R. J., Warford, J. J., & Wellenieus, B. (1994). *Telecommunications and Economic Development* (2nd edn.). Baltimore, MD: Johns Hopkins University Press.

Scifo, B. (2005). The domestication of camera-phone and MMS communication. In K. Nyíri (ed.), *A Sense of Place: The Global and the Local in Mobile Communication* (pp. 305–17). Vienna: Passagen Verlag.

Scott, J. C. (1998). *Seeing Like a State: How Certain Schemes to Improve the Human Condition have Failed.* New Haven, CT: Yale University Press.

Sellen, A. J., & Harper, R. (2002). *The Myth of the Paperless Office.* Cambridge, MA: MIT Press.

Sey, A. (2006). Mobile pay phone systems in Africa: the social shaping of a communication technology. Paper presented at the 56th Annual Conference of the International Communication Association, June 19–23, Dresden.

(2007). Mobile phones and the pursuit of sustainable livelihoods: why connectivity matters. Paper presented at the ICA Pre-conference Workshop "Mobile Communication: Bringing us together or Tearing Us Apart?", May 23–24, San Francisco.

Sharp, L. (1952). Case 5: steel axes for stone age Australians. In E. H. Spicer (ed.), *Human Problems in Technological Change: A Casebook* (pp. 446–60). New York: Russell Sage.

Silverstone, R., & Haddon, L. (1996). Design and domestication of information and communication technologies: technical change and everyday life. In R. Silverstone & R. Mansell (eds.), *Communication by Design: The Politics of Information and Communication Technologies* (pp. 44–74). Oxford: Oxford University Press.

Simmel, G. (1971). *Georg Simmel: On Individuality and Social Forms,* ed. D. N. Levine. Chicago: University of Chicago Press.

Singhal, A., Svenkerud, P. J., & Flydal, E. (2002). Multiple bottom lines?: Telenor's mobile telephony operations in Bangladesh. *Telektronikk* 1(2002), 153–60.

Smith, A., & Williams, K. D. (2004). RU There? Ostracism by cell phone text messages. *Group Dynamics: Theory, Research and Practice* 8(4), 291–301.

Smoreda, Z., & Thomas, F. (2001). Social networks and residential ICT adoption and use. Paper presented at the EURESCOM Summit 2001

3G Technologies and Applications conference, November 12–15, Heidelberg.

Sonenshine, T. (1997). Is everyone a journalist? (The impact of new technologies on newsgathering and the decision making processes of editors). *American Journalism Review* 19(8), 11–13.

Souter, D., Scott, N., Garforth, C., Jain, R., Mascararenhas, O., & McKemey, K. (2005). The economic impact of telecommunications on rural livelihoods and poverty reduction: a study of rural communities in India (Gujarat), Mozambique, and Tanzania. Retrieved November 1, 2007, from www.telafrica.org/R8347/files/pdfs/FinalReport.pdf.

Standage, T. (1998). *The Victorian Internet: The Remarkable Story of the Telegraph and the Nineteenth Century's Online Pioneers*. New York: Walker.

Steele, C., & Stein, A. (2002). Communications revolutions and international relations. In J. E. Allison (ed.), *Technology, Development, and Democracy: International Conflict and Cooperation in the Information Age* (pp. 25–53). Albany, NY: SUNY Press.

Steenson, M., & Donner, J. (2009). Beyond the personal and private: modes of mobile phone sharing in urban India. In S. W. Campbell & R. Ling (eds.), *Mobile Communication Research Annual* (Vol. I, pp. 231–50). Piscataway, NJ: Transaction.

Steinbock, D. (2003). *Wireless Horizon: Strategy and Competition in the Worldwide Mobile Marketplace*: New York: AMACOM Div. American Management Association.

Stevenson, W. B., & Bartunek, J. M. (1996). Power, interaction, position, and the generation of cultural agreement in organizations. *Human Relations* 49(1), 75–104.

Strayer, D. L., & Drews, F. A. (2004). Profiles in driver distraction: effects of cell phone conversations on younger and older drivers. *Human Factors* 46(4) (special section), 640–50.

Strayer, D. L., Drews, F. A., & Crouch, D. J. (2006). A comparison of the cell phone driver and the drunk driver. *Human Factors* 48(2), 381–92.

Strayer, D. L., Drews, F. A., & Johnston, W. A. (2003). Cell phone induced failures of visual attention during simulated driving. *Journal of Experimental Psychology: Applied* 9(1), 23–32.

Strayer, D. L., & Johnston, W. A. (2001). Driven to distraction: dual-task studies of simulated driving and conversing on a cellular telephone. *Psychological Science* 12(6), 462–6.

Su, D. (2005). The economy of "lucky" numbers: when old superstition needs new media. Paper presented at the International Conference on Mobile Communication and Asian Modernities II, October 20–21, Beijing.

Sun, H. (2006). Employee social assets, employer economic liability: cell phone usage and everyday resistance of live-in maids in Singapore. Paper presented at the International Communication Association conference, June 16, Dresden.

Sutherland, E. (2008). *Counting Mobile Phones, SIM Cards & Customers*. Wits, South Africa: LINK Centre, University of the Witwatersrand.

Tall, S. M. (2004). Senegalese émigrés: new information & communication technologies. *Review of African Political Economy 31*(99), 31–48.

Taylor, A. (2005). Phone-talk and local forms of subversion. In R. Ling & P. Pedersen (eds.), *Mobile Communications: The Re-negotiation of the Social Sphere* (pp. 149–66). London: Springer.

Thompson, E. C. (2005). Reaching home by hand-phone: foreign worker communities and mobile communication in Singapore. Paper presented at the Mobile Communication and Asian Modernities conference, June 7–8, Hong Kong.

Thorns, D. C. (1972). *Suburbia*. London: MacGibbon and Kee.

Tikkanen, T., & Junge, A. (2004). *Realisering av en visjon om et mobbefritt oppvekstmiljø for barn og unge* (No. RF – 2004/223). Stavanger: Rogalandsforskning.

Tönnies, F. (1963). Gemeinschaft and Gesellschaft. In T. Parsons, E. Shils, K. D. Noegele & J. R. Pitt (eds.), *Theories of Society* (pp. 191–201). New York: The Free Press.

Torero, M. (2000). *The Access and Welfare Impacts of Telecommunications Technology in Peru* (Center for Development Research Working Paper, No. 27). Bonn: Center for Development Research, Universität Bonn.

Townsend, A. M. (2000). Life in the real-time city: mobile telephones and urban metabolism. *Journal of Urban Technology 7*(2), 85–104.

Traugott, M., Joo, S.-H., Ling, R., & Qian, Y. (2006). *On the Move: The Role of Cellular Communication in American Life. Pohs Report on Mobile Communication*. Ann Arbor, MI: University of Michigan, Dept. of Communication Studies.

Trosby, F. (2004). SMS, the strange duckling of GSM. *Telektronikk 3*(2004), 187–94.

UNCTAD (2008). Information economy report 2007–2008: science and technology for development – the new paradigm of ICT. Available from https://unp.un.org/details.aspx?pid=17622.

Uy-Tioco, C. (2007). Overseas Filipino workers and text messaging: reinventing transnational mothering. *Continuum 21*(2), 253–65.

Vaage, O. (2006). *Mediabruks undersøkelse*. Oslo: Statistics Norway.

Varbanov, V. (2002). Bulgaria: mobile phones as post-communist cultural icons. In J. E. Katz & M. Aakhus (eds.), *Perpetual Contact: Mobile Communication, Private Talk, Public Performance* (pp. 126–36). Cambridge: Cambridge University Press.

Veeraraghavan, R., Yasodhar, N., & Toyama, K. (2007). Warana unwired: mobile phones replacing PCs in a rural sugarcane cooperative. Paper presented at the 2nd IEEE/ACM International Conference on Information and Communication Technologies and Development (ICTD2007), December 15–16, Bangalore, India.

Virilio, P. (1995). *Speed and Information: Cyberspace Alarm!* Amsterdam: Patrice Reimens.

Waverman, L., Meschi, M., & Fuss, M. (2005). The impact of telecoms on economic growth in developing nations. Retrieved November 1, 2007, from www.vodafone.com/etc/medialib/attachments/cr_downloads.Par.78351.File.tmp/GPP_SIM_paper_3.pdf.

Weber, M. (1958). *From Max Weber: Essays in Sociology.* New York: Oxford University Press.

(1997). *The Methodology of the Social Sciences.* New York: Free Press.

Wei, C. Y. (2007). Mobile hybridity: supporting personal and romantic relationships with mobile phones in digitally emergent spaces. Unpublished doctoral dissertation, University of Washington, Seattle.

Wei, R., & Lo, V.-H. (2006). Staying connected while on the move: cell phone use and social connectedness. *New Media and Society 8*(1), 53–72.

Weiner, E. (2007). Our cell phones, ourselves. *National Public Radio* Retrieved June 19, 2008, from www.npr.org/templates/story/story.php?storyId=17486953.

West, J. (2000). Institutional constraints in the initial deployment of cellular telephone service on three continents. In K. Jakobs (ed.), *Information Technology Standards and Standardization: A Global Perspective* (pp. 198–221). London: Idea Group Publishing.

White, R. (2006). *Swarming and the Social Dynamics of Group Violence* (Trends and Issues in Criminal Justice No. 326). Canberra: Australian Institute of Criminology.

Wong, A. (2007). The local ingenuity: maximizing livelihood through improvising current communication access technology. Paper presented at the Ethnograpic Praxis in Industry conference, Keystone, Colorado.

World Bank. (2007). *World Development Indicators 2007.* Washington, DC: The World Bank.

World Bank Global ICT Department. (2005). *Financing Information* and *Communication Infrastructure Needs in the Developing World: Public and Private Roles* (Working Paper No. 65). Washington, DC: The World Bank.

Yu, H. (2004). The power of thumbs: the politics of SMS in urban China. *Graduate Journal of Asia-Pacific Studies 2*(2), 30–43.

Zainudeen, A., Samarajiva, R., & Abeysuriya, A. (2006). *Telecom Use on a Shoestring: Strategic Use of Telecom Services by the Financially Constrained in South Asia* (No. WDR0604). Colombo, Sri Lanka: LIRNEasia.

Zainudeen, A., Sivapragasam, A., de Silva, H., Iqbal, T., & Ratnadiwakara, D. (2007). *Teleuse at the Bottom of the Pyramid: Findings from a Five-Country Study*. Colombo, Sri Lanka: LIRNEasia.

Zhao, Y. (2004). Between a world summit and a Chinese movie: visions of the "information society". *Gazette 66*(3–4), 275–80.

Index